MAN AND HiS TECHNOLOGY

MAN AND HIS TECHNOLOGY

Engineering Concepts Curriculum Project
State University of New York

McGraw-Hill Book Company

New York · St. Louis · San Francisco · Düsseldorf · Johannesburg
Kuala Lumpur · London · Mexico · Montreal · New Delhi · Panama
Rio de Janeiro · Singapore · Sydney · Toronto

This book was set in Times Roman by Typographic Sales, Inc., printed and bound by Von Hoffman Press, Inc. The designers were Mary Gaitskill and Ted Smith; the cover design and front matter were done by Dudley Thomas. The drawings were done by Typographic Sales, Inc., and RSR Associates. The editors for "The Man–Made World" were Marnie Hauff and Eugene Wheetley. The editors for "Man and His Technology" were Mary Gallagher Daraio, Jack Maisel, and Charles Wade. Joan Oppenheimer supervised the production.

ECCP has been supported primarily by grants from the National Science Foundation.

4567890 VHVH 798765

Library of Congress Cataloging in Publication Data

Engineering Concepts Curriculum Project.
    Man and his technology.

    "Essentially a reprint of chapters 2 through 9 of The man–made world, with chapter 1 representing certain new material."
    This book and a companion volume, Man and his technology, problems and issues, constitute the primary reference texts for courses offered at the State University of New York at Stony Brook.
    1. Engineering—Methodology. 2. Technology—Methodology. I. Title.
TA153.E57 1973   620'.001'8   72-13695
ISBN 0-07-019510-2

ECCP STAFF

Co–Directors

    Dr. E. E. David, Jr.,

    Dr. J. G. Truxal, State University of New York
                         at Stony Brook

Executive Director

    Dr. E. J. Piel, State University of New York
                         at Stony Brook

Staff Associate

    Dr. Thomas T. Liao, Polytechnic Institute of Brooklyn

Consultants

    Mr. John S. Barss, Andover, Massachusetts

    Mr. Edward G. Blakeway, Raleigh Schools, North Carolina

    Dr. Ludwig Braun, Polytechnic Institute of Brooklyn

    Mr. Henry Fraze, University of South Florida

    Dr. Lois Greenfield, University of Wisconsin

    Mr. Lester Hollinger, Glen Rock High School, Glen Rock, N.J.

    Mr. Richard King, Staples High School, Connecticut

    Mr. Dean Larsen, Jefferson County Schools, Colorado

    Mr. James J. McNeary, University of Wisconsin

    Dr. David Miller, National Academy of Engineering

    Dr. John Nickel, Wichita State University, Kansas

CONTENTS

Chapter 3   OPTIMIZATION

Chapter 4   MODELING

Chapter 5   SYSTEMS

Chapter 9   **MACHINES AND SYSTEMS FOR MEN**

LIST OF CONTRIBUTORS

Directors and Staff, ECCP

Dr. E. E. David, Jr.

Dr. J. G. Truxal
State University of New York at
Stony Brook (Director)

Dr. E. J. Piel
State University of New York at
Stony Brook (Executive Director)

Dr. Thomas T. Liao
State University of New York at
Stony Brook (Staff Associate)

Consultants

Mr. John S. Barss
Andover, Massachusetts

Mr. Edward G. Blakeway
Raleigh Schools, North Carolina

Dr. Ludwig Braun
State University of New York at
Stony Brook

Mr. Henry Fraze
University of South Florida

Dr. Lois Greenfield
University of Wisconsin

Mr. Lester Hollinger
Glen Rock High School,
Glen Rock, N.J.

Mr. Richard King
Staples High School, Connecticut

Mr. Dean Larsen
Jefferson County Schools, Colorado

Mr. James J. McNeary
University of Wisconsin

Dr. David Miller
Polytechnic Institute of Brooklyn

Dr. John Nickel
Wichita State University, Kansas

Contributors

Dr. Jack Alford
Harvey Mudd College

Professor Hugh Allen
Montclair State College

Mr. Edmund Anderson
Monroe Senior High School, Wisconsin

Dr. E. J. Angelo
Bell Telephone Laboratories

Dr. John L. Asmuth
University of Wisconsin

Mr. N. W. Badger
Garden City High School, New York

Dr. Euval S. Barrekette
IBM Research

Dr. D. L. Bitzer
University of Illinois

Professor William B. Blesser
Polytechnic Institute of Brooklyn

Dr. Joseph Bordogna
University of Pennsylvania

Dr. Manfred Brotherton
Morristown, N.J.

Professor A. E. Bryson
Stanford University

Mr. D. R. Coffman
James Caldwell High School, New Jersey

Dr. Thomas Earnshaw
Episcopal Academy, Philadelphia, Pa.

Mr. Jay Emmer
North Shore High School, New York

Dr. R. L. Garwin
International Business Machines Corp.

Professor J. Richard Goldgraben
Polytechnic Institute of Brooklyn

Dr. A. Jay Goldstein
Bell Telephone Laboratories

Dr. Newman A. Hall
Commission on Education
National Academy of Engineering

Dr. L. D. Harmon
Bell Telephone Laboratories

Professor W. H. Hayt
Purdue University

Mr. Charles Hellman
Bronx High School of Science, New York City

Professor D. A. Huffman
University of California at
Santa Cruz

Professor W. H. Huggins
Johns Hopkins University

Professor C. E. Ingalls
Cornell University

Mr. J. G. Johnson
Sidwell Friends School,
Washington, D.C.

Dean Edward Kopp
University of South Florida, Tampa

Mr. A. E. Korn
James Caldwell High School, New Jersey

Dean William A. Lynch
West Goshen, Conn.

Professor George Maler
University of Colorado

Dr. W. R. Marshall, Jr.
University of Wisconsin

Dr. John R. Pierce
California Institute of Technology

Mr. Robert Putman
Highline High School, Seattle, Wash.

Dr. James Reswick
Case Western Reserve University

Mr. Rolla Rissler
Aurora Schools, Colorado

Mr. G. I. Robertson
Bell Telephone Laboratories

Mr. B. A. Sachs
Brooklyn Technical High School, New York City

Dr. Samuel Schenberg
Board of Education, New York City

Mr. Robert Showers
East High School, Green Bay, Wisc.

Professor W. M. Siebert
Massachusetts Institute of Technology

Mr. M. Simpson
West Essex High School, New Jersey

Dr. Jeffrey D. Ullman
Bell Telephone Laboratories

Professor Andires van Dam
Brown University

Mr. R. A. Went
West Essex High School, New Jersey

Professor G. Brymer Williams
University of Michigan

Dr. E. E. Zajac
Bell Telephone Laboratories

PREFACE

This book is one of two companion volumes (the other entitled *Man and His Technology, Problems and Issues*) which have constituted the primary reference texts for courses offered at the State University of New York at Stony Brook, primarily for students not majoring in engineering or the physical sciences. The material contained in these two books has been supplemented by readings assigned from the current periodical literature, especially the references cited in the *Problems and Issues* volume.

In the summer of 1972, the Engineering Concepts Curriculum Project (ECCP) staff, with National Science Foundation support, ran a three-week workshop at which the experiences with this college course were described along with a discussion by all participants of the possible structures and features of a college course on the interaction between technology and man or society.

As a result of this workshop and of the student interest at Stony Brook, the decision was made to publish these two volumes. This volume is essentially a reprint of Chapters 2 through 9 of *The Man-made World,* with Chapter 1 representing certain new material. The *Problems and Issues* volume is a discussion of a series of sociotechnological problems (population, health services, accidents, energy, noise pollution, and so on).

In the Stony Brook offering of the course, the basic text is the *Problems and Issues* volume. As the class discussion encounters the basic concepts presented in this volume, the students are referred to the appropriate sections and chapters of this book. In other words, the series of social problems constitutes the central theme of the book. In each case, an attempt is made to determine the magnitude and nature of the prolbem, to ascertain why the problem has arisen, and finally to discover the alternatives available to society and government—alternatives or options for a positive course of action to ameliorate the problem.

Thus, the general goal of the course is careful and logical understanding of the problem and the resulting alternatives. To achieve this understanding, the concepts of modeling, optimization, dynamics, feedback, and stability are presented as they are needed and motivated by the problem.

Since the start of the ECCP, the staff has learned of a wide variety of college courses for both nonengineering and engineering students, in which many of the concepts and cases discussed in these two volumes are used. Indeed, the foundation technology course of the Open University in the United Kingdom is entitled "The Man-Made World"; it incorporates many of the approaches which appear also in the ECCP books. (The two activities have cooperated for the past three years and have, of course, freely exchanged ideas and materials.) We are convinced that the common goal of greater technological literacy can be achieved only if many courses are developed; we welcome the opportunity to exchange information with all groups.

This preface is followed by a reprint of the Preface from *The Man-made World,* since there the basic philosophy of the course is presented. To the acknowledgment paragraph of that Preface, we are pleased to add a statement emphasizing the unusual opportunity for creative academic activity provided at State University of New York at Stony Brook, headquarters for the ECCP work.

E. J. Piel
J. G. Truxal

PREFACE TO THE MAN-MADE WORLD

Throughout the writing of this book, three ideas have guided the authors.

1. *Learning should be fun.* There are well over a million engineers and scientists in the United States. Most of these men and women chose their careers because they enjoy understanding and creating something worthwhile. Learning how nature and the man-made world interact is like a never-ending game: each new bit of understanding leads to a host of new questions.

2. *Subject matter should be relevant.* Naturally, there are parts of science which are concerned with fundamental questions—the structure of the atom, the origin of the solar system, and the history of life. These are really the concerns of the theoretical scientist, seeking to advance human knowledge.

Most people, however, are directly concerned with technology which surrounds them and shapes their lives. This is the man-made world: the technology which has been created to improve men's lives. This is the transportation systems, the new buildings forming the modern city, the water supply and electrical energy systems, modern communication including television and the picturephone, and the automated factory or traffic-control system. It is also the pollution, the debasing of the earth, the unwise use of natural resources. If a technical course is to be *relevant,* it should develop those concepts which lead to an understanding of such use and misuse of technology and science.

3. *Science should be easy.* In our educational system beyond junior high school, technical studies usually have the reputation of difficulty. There is no logical explanation for this. Social science is inherently a much more difficult subject than the physical or life sciences. Social systems are more complex than systems with which technology and science are concerned. The largest computer ever built is much simpler than the justice system of even a small city.

ECCP

This book is a product of the Engineering Concepts Curriculum Project (ECCP). The Project started in 1965, when it became clear that fewer and fewer students were electing the physical sciences. This situation seemed particularly alarming when we heard on the national scene that the United States was entering an "age of technology." The computer alone promised changes which would make the Industrial Revolution seem a minor development.

Survival of our civilization depends in part upon our ability to adapt to these technological developments and to control the changes so that they remain in the realm within which society can adjust. Technology cannot be allowed to develop in such a way that groups of people are forced to withdraw from society. In a democratic country, such control can be realized only with public understanding of the nature, the capabilities, the limitations, and the trends of technology. Furthermore, the advance of technology cannot be understood without consideration of the interplay with social, economic, political, and psychological forces.

The problems we face as a nation are monumental. We not only need to understand technology, we must also anticipate its side effects. Change today is so rapid and the visible effects are often so delayed that the wrong decisions can lead to major problems before the effects can be reversed. Unfortunately, history emphasizes how difficult the probelms are.

Thus, the primary objective of the book is a first step toward technological literacy.

THE CONCEPTS

Technology (or applied science) embraces far too many concepts for coverage in any one book, no matter how brief the treatment of each. Modern technology has three primary parts:

Systems   Materials   Energy

Any of these three could have been selected; we rather arbitrarily chose the first: systems. In particular, we have focused on a field which is often called *information systems* by scientists and engineers. The title indicates the concern with understanding information, how it is communicated, how it is used to control systems, and how men and machines interact in a system.

Why select systems? There are two reasons. First, the computer seems to be the single, most significant development in modern technology. With the computer (and automation), man can be relieved of the drudgery of routine decision making. Second, the systems approach is increasingly important in economic, political, and social studies.

EDUCATIONAL OBJECTIVES

In the 1970's, any text or course should include a clear statement of educational objectives. What should the reader or student be able to do after the course that he could not do before? How can we measure the success of the program?

While ECCP has prepared specific educational objectives for teachers to evaluate their effectiveness, the true goal of the course is a start toward technological literacy. Several concepts are introduced in the following chapters: decision making, modeling, dynamics, feedback, stability, and logical design. We hope for an understanding of these concepts and then a familiarity with how a computer operates and what it can and cannot do.

For example, queueing is discussed in the first half of Chapter 3. After this study, we hope the reader will recognize queues as he moves through his normal life, will understand various ways the queues can be controlled, and will realize that certain queues are desirable while we should attempt to reduce others by appropriate economic or personal sacrifices. He should better understand the operation of modern urban traffic-control systems, for example.

Thus, two features found in many introductory science courses are relatively unimportant here. This is *not* a course on the scientific method. We are not emphasizing the methodology of the pure scientist. Second, mathematical problem solving is secondary to the process of logical thinking.

As a result, the classes where this course has been taught in the past have been characterized by much more discussion than usual, much less lecturing by the teacher. This book does not present a series of scientific facts which are really irrefutable by the student. Rather, the book presents a series of significant, current problems in which the concepts provide understanding.

ACKNOWLEDGMENTS

Over the five years of *The Man-Made World,* the Project has amassed an awesome list of people and organizations to whom we should express gratitude. Any such attempt at a complete listing would inevitably omit some of the many sources in education, industry, and government, and we can only list here a few.

The National Science Foundation has been the primary financial support for the Project throughout its history.

The National Academy of Engineering's Commission on Education (previously the Commission on Engineering Education) was the originator of the Project.

The Bell Telephone Laboratories and the Polytechnic Institute of Brooklyn have made major contributions in manpower and facilities, with significant contributions by IBM, Johns Hopkins University, M.I.T., Purdue University, American Machine and Foundry Co., Measurement Control Devices, and General Electric.

Finally, during the five years of development, more than 10,000 students have taken the course and, through their teachers, provided feedback which has been the primary guide in the transition from the early notes of 1966 to the present book.

E. E. David, Jr.
E. J. Piel
J. G. Truxal

June 21, 1970

Technology To Serve Man

1

I | GALVANI'S EXPERIMENT

The setting is September 16, 1786, at the University of Bologna in Italy (the world's oldest institution of higher learning, with its history dating back to the eleventh century). While a severe thunderstorm rages outside, a 49-year-old physician, Luigi Galvani, studies the anatomy of a skinned frog that he killed earlier in the day. Preparing to cut the carcass, he accidentally touches a nerve with the scissors he is holding. To his amazement, a leg muscle jerks violently.

Intensive experiments over the subsequent weeks fail to reveal the reasons for the muscle action. Then one day his wife, sitting in the lab watching him, notices that the same muscle action occurs whenever he touches the nerve with metal if an electrostatic generator across the room happens to be running. Still unsure of his observations, Galvani works 5 more years before publishing his results showing that muscle action results from electricity flowing through the nervous system.

Neither medicine nor electrical engineering were ready in 1791 to exploit this scientific information in any useful way. It is only in the past few years that engineers have used the information to design heart pacemakers that artificially pulse the heart regularly to cause beating of a muscle when the individual's own mechanism fails, or to design an artificial arm that can be moved naturally by the normal electrical signals from the patient's brain.

In the mere span of the past 15 years, the pacemaker has been implanted in the United States alone in more than 100,000 patients, who are now alive and leading normal lives because of this device. Indeed, by the spring of 1972, the pacemaker had become such a standard device that one company found a small group of its devices defective and (like the auto manufacturer) issued a recall order asking patients to come in and have their pacemakers replaced by an improved model.

The account above portrays the nature of technology today—the exciting potential for the rest of this century and also some of the problems. We find increasingly that technology has an impact on almost every aspect of our lives. We are living in an environment in which we seek a balance between nature and the man-made world—where we as a nation must bake basic social and political decisions on how to use technology to improve the quality of life for all the people.

Technology provides the opportunities to raise the general standard of living, to improve health, to augment mobility and the richness of life, and to develop better educational programs particularly individualized so that each person can learn at his own pace and in the directions he finds interesting. At the same time, increasing urbanization of the country and growing productivity, which requires fewer and fewer of the people working in agriculture or manufacturing industry, mean that the social problems associated with a technological age are steadily becoming more difficult to solve.

Thus, in this age of technology, we have both opportunity and danger:

Opportunity to improve the quality of life for all people

Danger that technology developed without public understanding will cause harmful side effects that will lead to serious degradation of the environment or social structure.

In this book, we seek a better understanding of the nature of modern technology and its impact on man. In the first chapter as background, we shall consider briefly two examples of modern technology: the heart pacemaker and the automated health examination center. Within these two cases, we shall find many of the problems and opportunities that continually reappear in the later chapters.

2 | HEART PACEMAKER

Electrical stimulation of human tissue was undoubtedly familiar to the coastal South Americans of ancient times through their contacts with electric eels while swimming. In 1774, there was a report in England of a 3-year-old child being revived by electric shock applied to her chest after she had "died" from a fall. In 1791, Galvani published the report of his famous experiments involving electrical stimulation of frogs' legs.

In modern times, a 1929 medical report told of the restoration of heartbeat by electrical shock. It was 1952, however, before another clear demonstration was carefully reported in the literature.

During the late 1950's, the cardiac, or heart, pacemaker was first developed—an electronic oscillator to deliver an electrical pulse to the heart at regular intervals (for example, 70 times per minute) in order to keep the heart beating at a regular rate in spite of malfunctioning of the normal, human excitation system. Today, just a little more than a decade later, well over 100,000 pacemakers have been installed; most of the patients who received pacemakers since 1963 are alive and leading useful lives. The use of the pacemaker is today such a routine medical operation, it receives little press attention even when performed on a Supreme Court Justice.

Use of the Pacemaker

The pacemaker is used in case of heart block—either total, corresponding to a stop of the heart beating, or partial, representing such a low beating rate that heart output is dangerously low. The heart block may occur because of malfunctioning of the electrical system which normally stimulates the pumping of the heart, disease of the heart, and shock from surgery or other causes.

The average adult has a heart rate about 70 pulses per minute, although athletes and others may have a much lower rate (down to 40). Under stress or physical exertion, the heart rate rises to 170 beats/minute (e.g., when an astronaut steps onto the lunar surface or a test pilot enters a particularly dangerous maneuver).

The beating of the heart can be stimulated by an electrical pulse (applied at the appropriate location) of an amplitude about 3.7 volts, and an energy in the pulse of about 100 microjoules.

A Succession of Pacemakers

The earliest pacemakers were external oscillators (outside the body), with wires leading from the oscillator through the patient's skin to the heart. During installation, the chest was opened to allow the doctor to implant the electrodes at the desired spot in the heart.

This early model had certain disadvantages. The patient had to carry the oscillator around, strapped to his body. Bathing was impractical. There was a tendency for infections at the point where the two wires entered the skin. A normal life was impossible.

It was fortunate for the millions of future users that during the 1950's there were extensive military research and development in the United States on miniaturization of components, long-life and reliable batteries, and electronic pulse circuits which require very little energy. As a direct result of this work, an implantable pacemaker became possible. The small oscillator circuit and battery (of a total size comparable to a package of cigarettes) were encased in plastic and placed under the skin on the left side at about waist level. Two wires led from the oscillator within the body up to the heart and were inserted with open chest surgery.

With a battery life of perhaps 16 months, patients had to return to the hospital regularly to have the battery replaced, but the operation was minor since it was only necessary to cut directly above the oscillator.

In 1965 the transvenous pacemaker was introduced. In this system, the oscillator (and battery) are implanted. The leads are then fed into a vein near the oscillator and pushed through the vein to the heart (just as a catheter is installed). The physician installing the pacemaker watches the wires moving through the veins toward the heart with a fluoroscope, possibly supplemented by X-ray pictures. Thus, only one operation is involved, which is at the point where the oscillator is implanted; the dangerous open chest surgery is avoided entirely.

Furthermore, the transvenous pacemaker proved especially attractive for emergency situations. The pacemaker leads can be inserted through the skin into a vein in the upper shoulder and pushed to the heart (with the oscillator-battery package left outside the body during the period of emergency care).

As always with technological development, these achievements failed to satisfy the user, in this case the physician. As soon as new technology becomes available to solve a particular problem, the customer always dreams of even greater potentialities. The next step in pacemaker evolution was in the direction of "noncompetitive" pacing: in other words, a pacemaker which pulses the heart *only* when the normal mechanism is inoperative. If the pacemaker is set for 860/1000 of a second between heart beats, the oscillator emits a pulse only if there is no natural pulse by the end of this time interval since the last beat.

There are two major advantages of such a system. First, if the patient's heart is beating normally, the pacemaker is idle and we can predict much longer life for the batteries. Actually, such devices have been used only for the past few years, and data on longevity are not available. It is possible that the longevity will actually be less than anticipated because of the more complex circuitry and the extra (third) wire needed to detect normal heart activity. The second advantage is of direct interest to the patient. By making the artificial pacemaker subsidiary to the normal functioning of the heart, there is possibility of a system in which heart rate increases automatically during periods of personal need—e.g., violent exercise. Thus, a more normal life is possible.*

The major disadvantage of the noncompetitive pacer is its susceptibility to electromagnetic interference. Since the electronic pulse generator operates in response to electrical signals from the heart, any stray electrical signals which pass through the skin to the oscillator and leads are apt to be detected as heart signals. In the presence of strong electromagnetic signals from external sources, the pacemaker may even stop operation.

Thus, in about 15 years, the electronic pacemaker has moved through an evolutionary cycle from a very crude device used only on critically ill patients in a few research hospitals, to a widely accepted prosthetic device, available even in small-town hospitals.

*
In order to allow patient control of the heart rate, one system includes a small, portable radio transmitter with which the patient can send signals through the skin to the oscillator to command a desired change in rate. When the patient is ready for strenuous exercise, he turns up the rate to perhaps 145 beats per minute; at the end of the exercise, he turns it down. The disadvantages of such a system are obvious: often we do not know in advance when we are to become excited.

Other possibilities for manual adjustment of the oscillator frequency are by an externally controlled electromagnet, or by a needle which is inserted through the skin to move the oscillator control mechanically.

As recently as 5 years ago, scientific meetings reported on the early experiments, in terms of the results achieved with 20 test patients. Today's meetings are concerned with discussions of which *type* of pacemaker is most appropriate for a particular medical problem; the pacemaker is a standard therapy for both heart block and many other irregularities in heart rhythm.

Cost-benefit Analysis

One of the important analytical tools widely used in business and government is cost-benefit analysis, involving a quantitative investigation of the dollar benefits of a particular program versus the cost. While such analysis involves intensive studies of the economic impacts following a particular innovation, we can easily make an initial and superficial evaluation of the pacemaker.

When 100,000 adult males have been kept alive and leading normal lives with the pacemaker, we can estimate the saving to all levels of government as $1 billion per year ($10,000 per man, representing the taxes he pays on earned income and the costs of support of his dependents had he died from heart block). One billion per year is approximately the income expected from an investment of $20 billion.

We are saying that even with these crude calculations and conservative estimates, the pacemaker justifies $20 billion of governmental investment during the last two decades in electronics research and development a justification which includes no dollar value placed on the human lives saved and the associated improvement in the quality of life in this country.

Furthermore, the pacemaker is a byproduct of our federal research program which could not have been anticipated in the early 1950's. (We recall from the beginning of this section that it was only in 1952 that the first modern case occurred in which heart beating was restarted by electrical pulsing.)

Technological Problems

The evolution of the pacemaker over the past decade has been marked by a series of technological problems. We mention a few to illustrate the engineering contributions that have been made by researchers, particularly in companies which have been involved in manufacturing the device.

1. Breakage of the flexible leads from the oscillator to the heart has caused some failures. The heart beats about once a second, and each beat flexes the leads.
2. Children present a particular problem since, as they grow, the leads must extend.
3. The plastic case containing the batteries and oscillator must remain intact in the body environment (that is, sur-

rounded by body fluids) and must be tolerated by the body. In the early years, particularly, the plastic casing tended to crack, especially near the exit of the leads.

4. There is a continuing search for better energy sources: batteries with life longer than the current 20 months, or implantable, nonchemical sources. In the latter category, piezoelectric sources have been tried: the mechanical motion of the heart itself or of the body leads to electric energy generation. A research program is studying nuclear energy sources, and in 1972 the first such pacemaker was implanted in a human being at the State University Center in Buffalo, New York.

5. In the realm of electronic circuitry, research focuses on reduction of the energy requirements, reliability, safeguards against the oscillator running away (e.g., with pulsing rates up to 300 per minute), and effective shielding from electromagnetic interference.

6. In 1972 trials started with a screw-in pacemaker which can be inserted through a very small incision in the chest. The hope is to avoid the difficult problem of locating the ends exactly right when the leads are pushed through a vein to the heart.

In each of these problems, reasonably satisfactory solutions have been achieved, and the refinements anticipated during the 1970's will provide to the physician a more reliable and flexible device with greater longevity.

3 | THOUGHTS ON THE PACEMAKER HISTORY

The very brief history given in the last section points out several important characteristics of modern technology. These are features which we find again and again as we look at the interaction of technology and man in the entire range of sociotechnological problems.

1. Once technology is available and has helped a group of people, the public in general tends to take it for granted. The extreme example of this phenomenon is the Apollo space program. The first moon walk was an event of worldwide significance. By the third moon walk, the public showed an awesome nonchalance and disinterest. In a less dramatic way, the pacemaker has rapidly gained acceptance as a very normal and unexciting device.

2. The popular opposition to the research efforts of the space and military programs of the past two decades seldom recognizes the extensive fall-out from such efforts. The

pacemaker is one example of devices which probably would not be available widely today if the United States had not made the tremendous investment in electronics research in the 1950's and 1960's. It is true that similar research could have been funded by the federal government for nonmilitary, nonspace purposes. Unfortunately, the recent history of the federal government indicates that major program funding is really only possible when the work has a goal which is publicly visible and politically attractive.

3. The pacemaker history is a dramatic example of the rapidity with which new science and technology are converted to public use. The rate of change is enormously accelerated today, even when compared with only a few decades ago. The typewriter was invented about 1870, but 30 to 50 years passed before use was widespread. The pacemaker moved from research to general use in little more than 5 years. Much more extensive studies of technological change have emphasized the steady increase in the speed with which such changes are occurring.

   This shorter time scale means that it is even more difficult for society and people to adjust to the changing environment. New technology tends to have a greater impact on people and their lives. Because change is so rapid, it is increasingly difficult to anticipate and control the side effects of new technology. As one example, after new food additives or drugs are introduced to the public, the growth in use is so rapid that very large numbers of people are affected, perhaps before serious side effects are evident.

4. The public desires are insatiable when new technology is involved. Once a workable pacemaker is available, the users want a series of improvements. As a result, the engineers are continually challenged to come up with better products—devices better matched to the human user.

5. Rapid technological change places extreme demands on other professional segments of society. In the pacemaker case, hospital administrators, nurses, physicians, and especially heart specialists have had to keep abreast of the year-by-year developments while continuing to fulfill their primary obligations to their patients. The greater the rate of technological change, the larger the stress on the professional worker, even though the working hours of the average citizen may be decreasing. Furthermore, the demands placed on the educational system are intensified.

6. Finally, as new technology affects more directly the lives of the citizens, effective consumer protection becomes increasingly important. In the pacemaker case, neither the physician

nor the customer is able to test the product of a particular manufacturer. National standards are essential to protect the lives of the customers.

The pacemaker is a specific device designed for a particular segment of the population. In the remainder of this chapter, we turn to a different form of technology-society interaction: the use of a variety of forms of modern technology in an attempt to ameliorate a social problem. Here we shall find that the problems and opportunities associated with the technology are somewhat different from the pacemaker situation.

4 | PREVENTIVE HEALTH CARE

The car owner typically takes his automobile into a service station or diagnostic center at least every 6 months for routine preventive maintenance. The hope is that such *preventive* care will avoid excessive bills and major failures at unexpected moments.

In congressional debates and other public forums, we frequently hear arguments for at least equal attention to the health of human beings. This country spends over $70 billion a year ($350 per person) on health care. Might a portion of this be saved and human suffering reduced by a formal program of preventive health care?

A major program to avoid illness and disability would include not only regular physical exams but also extensive educational efforts to teach nutrition and personal care, nationwide attacks on the problem of accidents, community programs to combat alcoholism, venereal disease, and lead poisoning, efforts to control harmful environmental pollution, and so forth. Certainly better health for Americans depends strongly on creation of an improved social environment in the cities and poor rural areas.

While regular physical exams are only one part of a preventive health care program, they are particularly interesting to us because in the past few years there has evolved here a strong trend toward the use of modern technology. Thus, in the remainder of this chapter, we want to look at the facilities which are arising for providing regular health examinations to a large segment of the population.

Screening

If we hope to improve health, one possible way is to measure frequently the health condition of large numbers of people. If disease or a serious condition is found, the individual is referred to a doctor for personal treatment. An example of this type of *mass screening* (testing of many people) is the mobile chest X-ray unit, which used to travel from location to location and take pictures by the thousands. The same type of mass screening is often found in elementary schools when pupils are tested for hearing, sight, or dental problems.

In the past few years, electronic equipment for testing a man's health condition has become much more elaborate. We can now test for many different diseases and malfunctions. We can often spot trouble before it is apparent to the man. Furthermore, in many cases the early detection makes medical treatment possible and simple.

At the same time, computers are now available that can accept very large amounts of data, remember and "file" these data, compare different test results, and then print out the data with a special mark beside the results which seem abnormal. The computer, in other words, allows the processing of quantities of data that simply could not be handled by human beings.

When we consider these two developments (better instruments and computers), the next step is evident. We can build a health-testing center: a building where thousands of people each year can receive health examinations. A computer collects all the data and automatically sends to each person's physician a health report on that individual. The building, or center, need have no doctors present. It is merely a testing center.

Such a building is called an *automated multiphasic health testing* center (abbreviated AMHT center). The word multiphasic means that many different phases, or aspects, are measured. If it were to take only chest X-rays, the center would not be multiphasic. It usually also measures hearing, sight, blood pressure—often as many as twenty different tests. Thus, the AMHT center gives an examination that covers a wide span of health. The word automated means that data are processed by the computer for the physician, special tests can be suggested, and unusual results are flagged or specially marked.

The Health Problem

In spite of all the publicity about heart transplants and medical research, there are strong indications that the health condition of the American people is not impressive. For example, life expectancy in the United States has not improved significantly during the past 20 years (Fig. 1–1). From 1959 to 1966, the United States dropped from 13th to 22nd place among the nations of the world in male life expectancy.

Even more dramatic is the low standing of the United States in infant mortality. In this country, this figure is now about $\frac{22}{1000}$. In other words, of each 1000 babies born alive, 22 will probably die during their first year. The corresponding figure in Sweden is about one-half as large (Fig. 1–2).

Fig. 1–1.
United States
life expectancy at
birth for
females.

What are the reasons for this standing of the United States?* Certainly many factors contribute. In a recent year in New York City, for example, 40% of the women who gave birth had not visited a doctor during their pregnancy. These people are separated from the medical and health-care system. Apparently they do not know how to obtain medical help. Many live a life separated from the familiar social services.

This separation between a large part of the population and the health-care system is emphasized by the infant mortality rates for different groups. In the major cities, for instance, the rate is about $\frac{35}{1000}$ compared with $\frac{22}{1000}$ for the entire United States. There has been a slight drop since the infant-care program of the late 1960's, but city people are still at a severe disadvantage.

The statistics are even more troubling if we look at the non-white population. Infant mortality is over $\frac{40}{1000}$ (almost twice the figure for the entire country). Clearly, this part of the population is separated from the health-care system. The separation is emphasized by the fact that only 2.2% of physicians are black, while 11% of the total population is black.

Thus, certainly a major part of the United States health problem arises from our failure to bring reasonable health care to a large fraction of the people. The American makes 5.5 visits per year to a doctor, compared with 3 visits by the people in Sweden (the usual standard for health care). This seems to indicate that we have an adequate supply of doctors. We simply do not use them properly.

*
A much more detailed discussion of this problem and its causes is given in Chap. 3 of the companion volume "Man and His Technology: Problems and Issues."

In region upon region of the central city, we find an alarming shortage of physicians (often perhaps one doctor for 25,000 people). The AHMT center offers the possibility of bringing this neglected population into the health-care system. Through a network of AMHT centers in the cities and mobile centers for rural areas, we can hope to bring reasonable health care to *all* the people.

History

The first AMHT facility was the Kaiser-Permanente Center established a decade ago in Oakland, California. This pioneering program is still the center regularly visited by anyone planning to start a new facility; it continues to experiment with new and improved methods of operation and to innovate in such areas as medical data processing and customer education.

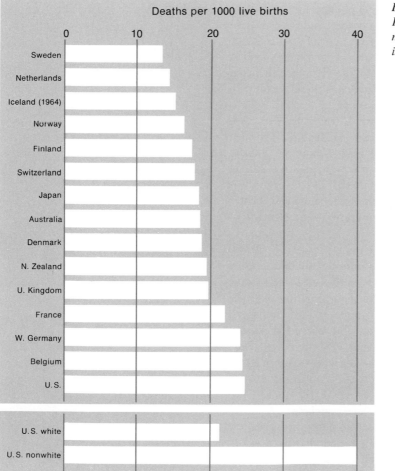

Fig. 1–2. Infant mortality rates in 1965.

In the late 1960's, there were several programs which built upon the Kaiser-Permanente experience. The State of California used a mobile center to test cannery workers, many of whom had never had reasonable medical care. The armed forces used modified centers for induction physicals. The U.S. Public Health Service established four pilot projects (Milwaukee, New Orleans, Providence, and Brooklyn) in order to determine suitable guidelines for groups which would later establish other centers. Several hospitals developed such centers to test entering patients.

More than 100 AHMT centers are now in operation, and many more are being planned or built. Some commentators on the health scene have expressed the hope that by another decade, 100 million people would be tested annually. If this size were reached, 5000 centers would be required since the average center handles about 20,000 customers a year. Such a growth would require a rapid increase in educational programs to train electronic and medical technicians just for these centers. In the present state of affairs, *all* the medical electronic technicians expected to receive baccalaureate degrees during the coming decade would not be able to staff the 5000 centers even if there were no other job needs for these people. Yet, if the cost of an exam at a center is $40, testing 100 million people at a total cost of $4 billion does not seem an unreasonable fraction of this country's total health bill (now $70 billion and likely to be $150 billion in a decade).

Realization of the above growth rate is unlikely, however, simply because of uncertainty among the medical and health profession as to the desirability of such mass screening. Should the expansion of AMHT centers be given high priority among the many needs of this country's health system? There certainly is a group of professionals who feel strongly that the AMHT center has not yet been proved to be a useful or important element of the total health system.

One argument is that a typical center, processing 20,000 customers a year, performs an enormous number of unnecessary tests. Such exams may be valuable when the customers have been separated from the health system for a long period of time; easily cured or arrested diseases are then found in large numbers. After these people have been tested once, however, the second tests will produce greatly diminished results. Furthermore, there are very few health difficulties which can be cured or arrested by early detection. Finally, mass testing leads to many false positive results (indication of trouble when there really is none). The result is a tremendous overburdening of the physicians.

We cannot determine whether such arguments are valid or not. Indeed, the medical profession seems to be sharply divided. Any sort of quantitative, rational decision requires studies over a long period of time of the costs of such screening and the effects on later disease and hospitalization. The Kaiser-Permanente Center is attempting to develop such a rational evaluation from the records of its customers over the years, but a long period of time and extensive research are needed before definitive answers are obtained.

Despite these reservations by part of the medical profession, the centers are expanding rapidly. The concept of group practice of physicians grew in spite of official objections until a few years ago, because both patients and many physicians like the group approach. Similarly, screening centers are now being operated or built by unions, organizations, health insurance companies, and industrial organizations hoping to make a profit.

5 | AN AMHT CENTER

To learn what an AMHT center is, we project ourselves a few years into the future and travel through a typical center. As we saw above, centers can be designed for many different types of customers: children, people chronically ill, pregnant women, and so forth. We choose a center which has been built to service a central-city area where the people have been separated from the health-care system. In other words, the primary objective is to provide basic health care needs for the customers.

First, the center is located well away from any hospitals or doctors' offices. It is truly a personal health center, and everything possible is done to make the visit pleasant. Two of us have made appointments together, so that we can talk as we travel from room to room.

As we pass through the front door, we enter a reception room. A pleasant hostess welcomes us and explains the purpose and

Fig. 1–3.
A possible patient flow schedule. In the text discussion, a different sequence is described. The actual sequence used depends on such factors as the tests to be given, the customers to be served and their attitudes and the desire to serve as many customers as possible in order to keep costs down. For example, if the desired customers are wary of the center and likely to leave before very many tests, it may be important to make the first few tests as familiar and pleasant as possible. (Searle Medidata Inc., a subsidiary of G. D. Searle & Co.).

PATIENT FLOW SCHEDULE
MEN | | **WOMEN**

7 MIN.
ADMISSION
Patient is identified to computer, issued an ID card and given a drink of glucose (if appropriate).

15 MIN.–30 MIN.
MEDICAL HISTORY STATION
Patient completes automated interactive medical questionnaire assisted by technician where necessary.

10 MIN.
VISION STATION
Visual acuity, phoria, depth perception, and color perception are tested.

6 MIN.
HEARING STATION
Audio acuity is measured, either self-administered or by a technician.

3 MIN.
UNDRESS
Patient removes upper garments, dons a smock.

3 MIN.
X-RAY STATION
Technologist takes chest X-ray.

5 MIN.
MAMMOGRAPHY STATION
Technologist takes breast X-rays (low-voltage, fine-grain film).

15 MIN.
CARDIOVASCULAR STATION
A 12-lead ECG is recorded. Pulse rate, blood pressure, and ocular tension are measured. Blood is drawn one hour after glucose ingestion.

5 MIN.
ANTHROPOMETRY STATION
Height, weight, and skinfold thickness are measured. Pulmonary function is tested by spirometry.

10 MIN.
CYTOLOGY STATION
Pap smear taken for cytology. Self-palpation taught.

5 MIN.
DRESS
Patient removes smock, dresses.

3 MIN.
LAVATORY
Urine specimen collected and given to technologist.

5 MIN.
DISCHARGE
Admission receptionist checks that tests are complete; directs patient to any optional tests (retinal photo, etc.) Discharges patient.

MEN | | **WOMEN**
77–92 MIN. | | 92–107 MIN.

procedures since we have not been there before. She hands each of us a personal identification card which we will show in each room to be sure the test results are credited to the right person in the computer. We are then led to chairs before a TV set, where we see a 7-minute program describing what we will be doing during the next 2 hours. As soon as the program ends, we are asked to go to Room 1.

In Room 1, an operator measures height, weight, and important body dimensions. All results are fed immediately into the computer, which from now on keeps a record of all test results for each of us. Very often, there is also a measure of obesity. This is done by measuring the thickness of excess fat in the upper arm. If the computer decides that we are too fat, we will receive a booklet or a monthly magazine on the danger of obesity and ways to lose weight.

In Room 2, hearing is checked. How this is done emphasizes the importance of matching the technology to the people using it. The standard way to give a hearing test is to put earphones on the patient. A signal at one frequency or pitch is then sent in through one earphone, with the loudness slowly increased. The patient indicates when he just hears the note. This procedure is repeated at several different frequencies in each ear. Thus, the sensitivity of each ear is measured at various frequencies.

When testing customers who are not accustomed to physical examinations and who are fearful of technology, it may be entirely wrong to use the earphones and put the audio equipment in the same room. One may find that a significant fraction of the customers at this point simply walk out of the testing center. If this occurs, the hearing test can be conducted by putting the audio equipment in an adjacent room. The signal is fed through speakers in the ceiling, and the customer sits in a comfortable chair at a set location. While the tester may not be able to determine accurately the response of each ear separately, he can obtain a measure of the customer's total hearing ability. And the testing can contribute to his relaxation and enjoyment.

Here again much more research is needed on what hearing tests to make. Recent studies have suggested that hearing ability is perhaps better measured by man's ability to understand spoken words in a background of loud noise than by the traditional, single-frequency test. As AMHT centers become commonplace, it is natural to ask again: what measurements do we really want for health assessment?

One true story of a screening center illustrates some of the unexpected problems that can arise in testing. In this center, an unusually large fraction of the customers showed hearing problems, in many cases even though the people themselves insisted that their hearing was fine. Instruments were rechecked and tests

*Fig. 1–4.
Hearing is
tested with an
automated audiometer.
(Kaiser Foundation
Hospitals)*

run on weekends to find out the reason for the peculiar results—all to no avail. Finally one day a customer complained that she could not hear the audio signal because someone had flushed a toilet (the rest room happened to be next door). When the hearing test was moved to another part of the building, results returned to normal.

Room 3 houses the eye tests. We read the usual eye chart. The field of vision is measured automatically—a computer moves a flashing spot around, and we indicate when we can see the flash. Finally, the pressure inside the eye is determined. (Particularly in middle-aged people, an increase in this pressure is a warning of glaucoma, a disease which inevitably causes blindness if it is not treated.)

Another room is devoted to respirometry, the measurement of the amount of air in each breath. We exhale into a balloon or similar device. The size of the balloon after one breath is measured. Finally, the data are fed to a computer which then decides whether a person is breathing normally on the basis of age, size, and sex. This test illustrates how essential patient cooperation is if results are to be meaningful. Obviously, if the customer exhales slightly, the test will really prove nothing.

As we travel through the series of testing rooms, measurements are made of blood pressure, heart rate, and the electrical signals causing the heart to beat (electrocardiogram). Blood and urine samples are taken, and chemical analysis is done by automatic equipment which sends the results directly to the computer. This automatic chemical analysis is very important in keeping the cost of the tests low. It is only in the last decade that we have had such equipment that can determine the amounts of twenty different chemicals in the blood.

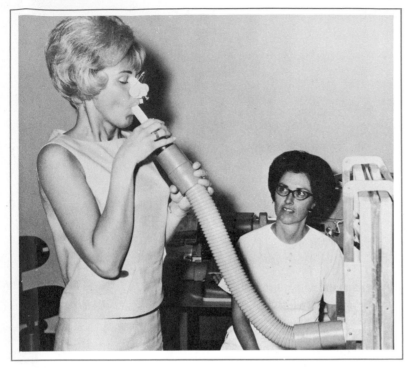

Fig. 1–5.
Vital capacity of
lungs is measured
by blowing into a
respirometer.
(Kaiser
Foundation
Hospitals)

In this way, perhaps twenty different sets of tests are made to check for a variety of problems—heart malfunction, venereal disease, breast cancer in women, and so forth. In each case, test results (in numbers) are sent to the computer, where they are combined into a total picture of the health of the patient. In each room, there are signs and short TV programs explaining what is being done, and attendants are always ready to discuss the tests in the native language of the customer.

In the last room, we answer questions from the history questionnaire. This is a series of maybe 300 questions regarding family background, habits such as smoking and drinking, a history of diseases and medical problems, and any recent symptoms such as headaches or dizziness, for example. The questionnaire takes about as long as all the other tests together. The information from this questionnaire, however, is of vital importance to the physician. In many cases, the patient is able to describe his problems much better than any measurements can reveal.

At the conclusion of the testing procedure, the patient is told his results will be sent directly to his personal physician within a few days. An appointment is made for the patient to visit his physician. If the patient has no regular doctor, he and the receptionist select one from a list of the doctors who call regularly at a clinic associated with the center.

Under no conditions are the test results given directly to the patient. Some of the results may be in error or cause undue concern. They must certainly be interpreted by a doctor before the patient is alarmed by some result that seems abnormal. If any results do appear dangerous, the doctor will undoubtedly want to have additional tests made before treating the patient.*

Thus, the AMHT center is a collection of sensitive instruments and a computer. The center is run by nurses, equipment operators, and technicians. These staff members use the equipment to make a complete survey of the patient's health for the physician. By using mass testing methods, the examination costs less than half as much as when done on an individual basis. Furthermore, in a large testing program, equipment of the best quality can be used and operators can be especially trained. Most important, the AMHT center can bring health care to the people. The center can be located in the middle of the population it serves, it can be staffed by people recruited from this same population, and testing at the center can be a pleasant and attractive experience.

6 | PROBLEMS WITH THE AMHT CENTER

There are serious problems in building such an AMHT center. These are problems which appear whenever technology is used in a social problem. Following is a discussion of three important problems.

*Fig. 1–6.
Electrocardiogram
(EKG) of heart is taken
and recorded on patient's
data-processing card.
(Kaiser Foundation
Hospitals)*

*
Actually there are some AMHT centers now in operation with a physician available for an interview to answer questions and discuss any test results which seem unusual. (In many centers, the blood and some other tests are not completed until a day later, mainly because of the high cost of immediate analysis.) Most opinion, however, seems to favor a later visit to a physician; any painful or uncomfortable tests such as sigmoidoscopy (rectal examination) can then be performed.

Fig. 1–7. A 70-mm X-ray is taken for careful study by a radiologist. (Kaiser Foundation Hospitals)

Measuring the Data

The first problem is largely scientific: What data should be measured and which tests should be used?

There are hundreds of tests which indicate a man's health. We might measure the strength with which he can grip a ball or measure reaction time in various tasks. We can measure a man's body temperature, his heart rate after different amounts of exercise, and the condition of his teeth.

In any system designed to help a social problem area, there is a tendency to measure everything. If we are not to be swamped by data, we must measure only data which meet four conditions:

1. *There must be instruments* which will make the measurements without unreasonable discomfort to the customer and at an acceptable cost. These limitations are severe. For example, in measuring sensitivity to pain, a small hammer is pressed against the Achilles tendon in the heel to determine the pressure when the customer moves or objects. Obviously, this is very crude way to measure pain sensitivity. If the customer is stubborn, he may allow severe pain before moving; another customer may jump at the mere touch of the cold hammer. The problem arises because we really do not want to inflict severe pain.

As another example, it would be very advantageous to measure malnutrition, particularly in pregnant women. Unfortunately, the only tests showing malnutrition which might affect the child but which is not severe enough to be obvious in the woman cost several hundred dollars—far too much for a mass screening program.

2. *The data must be meaningful.* They must be reasonably accurate. A false-positive report of trouble is often followed by a second test. If that is negative, the physician then faces a 1-1 score. As more tests are required, both doctor and patient rapidly lose confidence in the center. Furthermore, the data should be repeatable. One of the problems with measuring blood pressure by a sphygmomanometer (the air pressure tube wrapped around the upper arm) is that the reading depends on both the state of mind of the person doing the measurement and the nervousness of the patient.

3. *The data must concern treatable diseases.* One of the troublesome possibilities in current research is that it may be possible to predict heart attacks from electrocardiograms. If this happens, testing a group of middle-aged males may reveal that there are 1000 for each of whom the tests indicate that there is a 50% probability that he will have a heart attack within the next 6 months. With this capability, what does the doctor now tell each of these 1000? He doesn't know what to advise to avert a heart attack. He will needlessly terrify 500 men. The question then arises: is this information, acquired from the testing, of any real value, or does it do more harm than good?

4. *The total data must not make the doctor's job impossible.* One center measured 1800 facts about the patient. The typewritten description filled six single-spaced pages. In such a case, the doctor would probably rather give a new health examination than try to figure out what the computer output meant.

To meet these conditions, the usual AMHT center has at most about twenty different tests. Since each test may give several different numerical results, even this program results in a computer portrait of the patient's health which fills several pages. To help the doctor read this material, the computer "flags," or marks distinctly, the results that seem unusual. Often the first page contains only the abnormal data.

Thus, the design of the AMHT center requires careful selection of the tests to be given. The decision must depend on what technology is available, what knowledge is useful medically, and how much data can be profitably used.

Fig. 1-8.
*Automating
the analysis of
blood samples and
other body fluids has
contributed enormously to
the feasibility of multiphasic
health screening. Once tediously
and expensively done by hand,
analysis of blood constituents can now
be done by machine. Tests on great
numbers of people are thus economically
feasible. The machine's rate of
operation is 60 blood samples per hour
or 720 individual tests. Results are
recorded on a chart in the form
of a profile. Gray regions
indicate the normal range.
(Caption from* Machine
Design. *Photo from
Technicon Corporation.
SMA is a
registered trademark
of Technicon
Corporation.)*

SMA 12/60

Public Acceptance of an AMHT Center

The second major problem is: How do we gain acceptance of the technology by the public for whom it is built? In the case of the AMHT center, we emphasized that the center is for a part of the population which has been separated from the health-care system. These are people who are probably not highly educated, may be unemployed, and are not familiar with most modern technology. Furthermore, they are not familiar with the present health-care system.

Several principles have emerged from experience with the early AMHT centers. While these ideas may seem obvious now, they were often not recognized when the first centers were built.

1. The center should be located in the community it is to serve. Transportation is often very difficult for people in the central city. If the center serves rural people, or groups widely separated, it should be mobile (for example, in a trailer).
2. Going through the center should be a pleasant experience. Enough staff should be present to answer questions and explain the center. Instructions must be clear.
3. Persons employed in the center should be from the same ethnic group as the patients.
4. Signs and questionnaires must be in a language familiar to the customer. If this is Spanish, for example, the correct Spanish dialect should be used. In one center, the questionnaire was translated into Spanish by a noted professor. Unfortunately, he used the finest Castilian Spanish, which was as unintelligible as English to the clientele.
5. The questionnaire must be at an appropriate literacy level and must not ask for information which the patient cannot be expected to know. One proposed questionnaire contained the question:

    Have you had recurrent abdominal pains?　　Yes——
No——

Obviously many people would have no idea what the question meant.

Another problem with such questions is whether to include a "Don't know" answer in addition to "Yes" and "No." If don't know is included, some people soon learn that a don't know answer is an easy way to move to the next question and finish the work in a hurry. We may find 140 consecutive don't know responses. On the other hand, there are times when the patient really does not know. Many adults do not know whether they have ever had mumps. If we have only yes and no answers, how are they to respond?

6. The instruments used for testing must either be hidden or be attractive to the patient who is afraid of complex equipment. The success of many tests depends on the patient's cooperation. Other tests (heart rate, blood pressure) require that the patient be relaxed if the doctor is later to draw any useful conclusions from the data.

Problems occur again and again when we try to use technology to improve social conditions. The AMHT center is a product of modern technology. It is one small step forward in trying to bring benefits to all the population. In this sense, it is typical of the way technology can meet social needs. Yet it is only successful if we make sure the center is built for those people it is supposed to help.

The Effect of the AMHT Center on the Total Health System

A third major problem is whether the AMHT centers can be placed in the total health-care system. At the moment, the United States has a total system which includes doctors, nurses, hospitals, nursing homes, and so forth. This system is providing excellent health care to a sizeable portion of the population.

If we add several thousand AMHT centers in a very few years, can the other parts of the system adapt? Doctors are already very busy. If we test millions of people who have not been using doctors, can we save enough of the doctor's time with regular patients to take care of this new group?

If millions of people are tested, we can expect tens of thousands to be referred to ophthalmologists (eye specialists) for eye problems they had not been aware of. Are there enough specialists in each category?

Hospitals are already overcrowded in many cities. What happens if particular treatment is needed rapidly for large numbers of people each year?

In other words, new technology can make major changes in a total system. In order to provide the best care for the most people, the other parts of the system must adapt to these changes. Otherwise, we are apt to find that the new technology has unexpected side effects. If we do not plan carefully, the net result may be a final system which is poorer than the one we started with.

The problem is particularly serious because of the speed with which change occurs. The first AMHT center (Kaiser) was partially automated in about 1965. By 1969, a half dozen other centers were in operation. By 1970, 400 additional centers were planned. There is little time in which to build new hospitals or train technicians and operators. To educate a new generation of doctors or even to increase the number of doctors significantly requires additional time.

7 | TECHNOLOGY ASSESSMENT

In the last section, we discussed briefly some of the problems associated with the concept of the AMHT center. There are questions like

What data should be collected?

How can the design be matched to the target public?

How can we ensure compatibility with the rest of the health-care system?

Such concerns are the basis for the differences of opinion about AMHT among medical and health professionals. Quite naturally, each physician views the AMHT concept in terms of his own background and experience, the nature of his own practice, and his personal evaluation of the overall health problems.

When government officials have to make a decision about whether or not to expand the AMHT centers (or any new technology), they are confronted by another question which is really even more far-reaching: What will be the long-term impact of this innovation on the total quality of life? In other words, they should consider not only the immediate and direct effects on health care, but also the long-term influence on related economic and social systems. Sensible evaluation of these secondary, but often extremely important impacts of new technology, is an exceedingly difficult task, perhaps in many cases impossible.

Congressman Emil Daddario, who did not run for reelection in 1970 in order to pursue a campaign for governor of Connecticut, popularized the term *technology assessment* (or TA) for such a study. That is, technology assessment involves the anticipation, before a new technology is introduced, of the broad effects of the innovation in order to control the speed and manner of that innovation so that serious side effects can be avoided. During the past few years, there have been extensive efforts both in the federal government and in research centers to develop a strategy to carry out technology assessment.

Example of Goals of TA

In order to appreciate the difficulty of intelligent technology assessment, we can consult history to observe some of the side effects of new technology which were not (and perhaps could not have been) anticipated.

A classic example is the broad history leading up to the polio vaccine. About 1820, central town water supplies were introduced as a major technological innovation in England and the United States, as a consequence of the development of efficient pumps. By about 1870, typhoid epidemics were common and serious;

scientific studies led to the conclusion that central water supplies had to be chemically purified, and typhoid was brought under control.

Purification not only removed typhoid as a threat but also resulted in another change. Small children no longer normally had mild polio cases which resulted in an immunity. By the early part of this century, severe polio epidemics had become serious and frequent, with an accompanying heavy toll in lives and disabilities.

In the 1950's polio vaccines were introduced, and polio is controlled. In the light of the history, one cannot help but wonder whether the end of this century will bring to light a totally unanticipated side effect of the polio vaccine.

Had technology assessment been possible at the various stages of the above history, we would have anticipated the difficulties and could perhaps have averted at least part of the human tragedy. Unfortunately, technology assessment requires both an understanding of the full implications of new technology and an ability to predict the future, at least in part.

Other examples can be cited. In the early 1900's, the automobile was hailed as the salvation of the environment; no longer would city streets be covered with horse urine and manure. The effects on air pollution could have been anticipated only if we could have predicted the number of cars on the road today. Prediction of the disastrous effects on the central city through suburbanization would have required an awesome prescience.

In a similar vein, could we have predicted the role of the typewriter in creating jobs for women and liberating them from the home, or the revolution in health, life expectancy, population, and urbanization resulting in Latin America from the introduction of the window screen?

Importance of TA

Even though technology assessment is difficult at best, and maybe impossible, this country clearly must attempt it if we are to have any chance to *control* the way that technology develops. The slogan of today's world seems to be: the improbable is easy, the impossible takes a little while. Perhaps in the near future we will be able to look back on the history of technology assessment and see that it too falls within this slogan.

In the AMHT concept, technology assessment certainly encompasses the apparent problem: how do we ensure that the growth of centers does indeed benefit the segment of the population which is not now receiving adequate health care? From another viewpoint, how can we be sure that the AMHT centers are not primarily used by the suburban middle class as an adjunct to their already high-quality health care?

There certainly are governmental strategies which can avoid such difficulties. The development and acceptance of such strategies require an awareness of the potential dangers—the goal of technology assessment. Having recognized the problem, we can move to guarantee that centers are desirably located, that the tests are chosen for the target population, that the educational system is encouraged to produce the required, trained personnel, and that all young people are educated to understand the role of the technology.

Thus, any major technological innovation must be introduced with an appropriate, total-systems approach built on the best possible understanding of the likely impacts of the change. When this is the accepted strategy, we will be ready to ensure that the age of technology does lead to an ever-improving quality of life.

STEVENSON

"*Have you gone nuts? One hundred and fifty computer-programmed plays to call in the book, and you say, 'Everybody run downfield and I'll heave one.'*"

Decision Making

2

You have just been given permission to drive the family car, but your father keeps the key to the locked gas cap. He tells you that he will fill the tank on Friday night and you can have the car all day Saturday, but to drive slowly because the slower you drive, the better your gas mileage will be. You have a single decision to make: Drive slowly and get maximum mileage or drive fast and get minimum mileage. Or is it that simple?

The Committee for Clean Air of a large city meets with the mayor and demands that all incinerators used for burning garbage within the city limits be shut down immediately to cut down air pollution. The mayor promises to discuss the problem with his commissioners of traffic, sanitation, and rivers and harbors, and report back to the committee. The committee chairman snorts, "More bureaucracy," and marches out of the meeting to a press conference where he reports that the mayor is stalling on a simple question which has a very simple answer. The mayor in return replies, "For every complicated question there is an answer which is forthright, simple, direct, and *wrong!*"

In each of these cases, there are many decisions to be made—some insignificant and others crucial. In this chapter, we look at the nature of decision making and at some of the techniques which can be used to arrive at intelligent decisions.

As an example, let us look at the first "simple" situation described at the beginning of this chapter. This situation contains elements common to all decisions. How fast to drive a car is a simple question. The rule is not, Drive slowly and get maximum mileage, or drive fast and get minimum mileage. We have to look at the problem more carefully.

Figure 2–1 shows how gas mileage depends on speed for a particular car. At a standstill, with the engine running, the effi-

Fig. 2–1. Gasoline economy as a function of automobile speed.

ciency is zero—no matter how much fuel we use, the automobile doesn't get anywhere. As the speed increases, the gas economy improves, until at 45 miles per hour, it reaches a maximum of twenty-two miles per gallon. Beyond 45 mph, the economy falls off because of friction in the engine, in the wheel bearings, and in the air through which the car moves, as well as because of a decrease in engine efficiency at high speeds. It is obvious that the most economical operation of this car is realized when we drive at a constant speed of 45 miles per hour.

It is rarely possible, in any realistic situation, to travel at a constant speed, because of traffic conditions, curves, hills, and weather conditions. Fortunately, in this example, small departures from 45 mph do not cause a serious problem. At all speeds between 35 mph and 55 mph, the gasoline economy is within 5% of the maximum. As long as we stay reasonably close to 45 mph, we are operating nearly at an optimum condition.

We base our decision to drive at 45 mph on our desire to maximize the gasoline economy (minimize the cost of gasoline per mile). In the real world, decisions seldom are so easily made. There often are many factors which we must consider. Also, many times we are not sure what is really best (or optimum).

If we were to drive through the center of a city, where the speed limit is set at 25 mph, the police would take a very unsympathetic attitude toward a speed of 45 mph. Our decision making becomes somewhat more complex because of this constraint. Its effect is shown in Fig. 2–2. The legally imposed constraint lets us drive our car at any speed between zero and 25 mph. If we wish to maintain maximum gasoline economy *under this constraint*, we must operate at 25 mph. This is the highest point on the curve of gasoline economy versus speed within the allowed region.

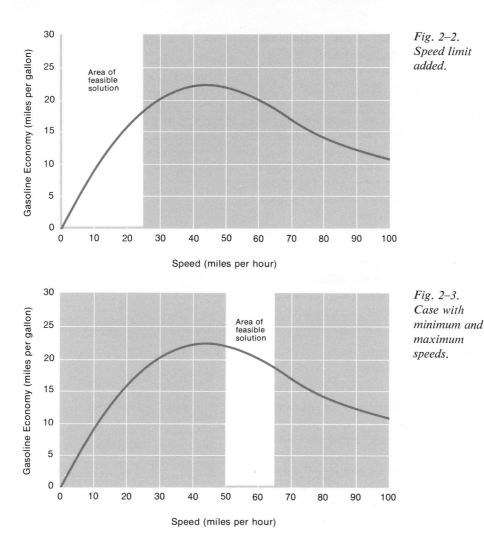

Fig. 2–2.
Speed limit
added.

Fig. 2–3.
Case with
minimum and
maximum
speeds.

If, on the other hand, we drive along a limited-access road such as the New York Thruway or the Pennsylvania Turnpike, a different set of constraints may be imposed. One such set is shown in Fig. 2–3. Here, we are not permitted to travel faster than 65 mph, nor slower than 50 mph. We should note here that if our minimum-speed constraint had been 40 mph rather than 50 mph, the decision would have been to drive at 45 mph rather than at either limit. We want the *maximum* possible economy.

Frequently, we are required to make decisions on the basis of conflicting desires. In our limited-access-road example of Fig. 2–3, we may be required to travel between two points in as short a time as possible. Perhaps we have a date with a girl who becomes angry if we arrive late. In addition, we may wish to use as little fuel as possible. We see from Fig. 2–3 that we cannot simultaneously drive at maximum speed (within the speed limit, of course) and maximum economy. The two goals just do not go together. Such conflicts are typical of real decisions.

We may decide to drive at 50 mph for economy and to risk the wrath of our girl friend. If we do not wish to assume this risk, we may drive at 65 mph and arrive on time to find a happy young lady.

As an alternative, we may compromise and choose some speed between 50 mph and 65 mph, perhaps 58 mph. This compromise lets us save some money and not be too late. Deciding on such a compromise really means we have chosen a new goal. We have decided that the few minutes we save are worth the cost of the extra gas.

Unfortunately, most decisions involve such a compromise. We have to decide on the relative importance of the different goals.

The speed example contains all of the parts of a decision-making problem. Let us list these so we can return to them in later examples. There are four elements:

Model   Constraints
Criteria   Optimization

The definitions of these terms are:

1. *Model*. The model is the mathematical or quantitative description of the problem. In our example, we are interested in obtaining the maximum gasoline economy; hence the model is the curve of Fig. 2–1, which shows how gasoline economy depends on the speed. In the following chapters we consider the various forms of models in much greater detail, since the model is the item which changes the problem from one of intuition or common sense into a quantitative problem which we can hope to solve precisely.

2. *Criteria*. The criteria are the goals or objectives of the decision-making problem. In our example, the criteria are to achieve the best possible gasoline economy and to arrive at our destination in as short a time as possible. When there are several criteria which are incompatible, we have to select a criterion which is a compromise among these.

One of the most dramatic decision problems occurs when there is a major catastrophe (fire, explosion, earthquake, or battle) with a very limited number of doctors available. The problem is to decide how to use the doctors in the best way. The injured people are then sorted into three groups: those who cannot be saved, those who can probably be saved by a doctor, and those who will survive without immediate attention by a doctor. The decision problem is then normally solved by the doctors working on only the middle group.

3. *Constraints.* Constraints are added factors which must be considered in the solution of the decision problem. In our speed example, the speed limits represent constraints. Other constraints in our example would be unsafe road conditions and driver fatigue. (Speed may be limited by road conditions, and the time involved in long trips may have to include rest periods.) Thus, the constraints specify the region within which we should look for a decision. We must find the best solution which satisfies the constraints.

4. *Optimization.* Once the problem is formulated (the model), we decide what we really want (the criteria), and we know what is permissible (the constraints), we are ready to try to find the best, or *optimum*, solution. In our speed problem, we found a solution by merely examining the model and considering the constraints. In more complex problems, it may be necessary to find special engineering or mathematical techniques; in many practical cases, we have to use a trial-and-error approach.

In this section we have considered a very simple example of a decision problem (the selection of the optimum speed at which to drive a car). We found all four elements of the typical decision-making problem: model, criteria, constraints, optimization. Next, we want to show a few other decision problems with different forms for the model, criteria, and constraints. We look also at a few different ways to optimize. Finally, we want to suggest the wide variety of decision problems.

2 | TYPES OF DECISIONS

In the preceding section, we found that the speed example of a decision problem has three parts. The model describes the situation. The constraints say what we cannot do. The criterion is the basis for a decision. Once these three parts are known, we optimize. That is, we find the *best* solution.

It is amazing that so many different decision problems can be viewed in this way. In very simple problems, nothing is gained by trying to find the model, constraints, and criteria. A man is quite good at making simple decisions without worrying about how he does it. For example, in a restaurant he orders according to the prices and what he feels like eating. Only a crazy scientist would think that the menu is part of the model. The rest of the model is how well he likes each food. The constraint is the amount of money he is willing to spend. The criterion is to find the best combination of foods which costs less than the limit. Obviously, if we had to think about all this, we would die of starvation. But there are many problems where we can gain much better understanding if we recognize these parts. The two examples which follow should provide some additional insight.

*Fig. 2–4.
Which road
to take?*

Choosing a Road

We are driving into the center of a city. We reach a "V" in the road. Here we can reach the city by taking either route, and we have to make a decision: right or left. Figure 2–4 shows the problem. Our only criterion is to reach the city center as soon as possible.

If the speed limit is the same on each road and if there is no traffic, the decision is simple. However, we have driven both ways before. We know that there is always traffic, usually heavier on the shorter route. Furthermore, any accident ahead on either road is likely to cause enormous delays. How do we decide?

Obviously in a problem like this, we cannot make an intelligent decision. We just don't know the model. Is there an accident today? If so, where? What is the speed of traffic five miles ahead on each road? How long will it probably take on each road? We cannot make an intelligent decision without knowing the model, at least approximately.

This example emphasizes one of the simple improvements in traffic systems which could be made with relatively little money. A thruway in a major city may cost as much as $28,000,000 per mile to build. For a very small fraction of this amount, we could measure the average traffic speed at several points. This information could be sent back by telephone wires to the driver's decision point. As we approach the "V" of Fig. 2–4, we would see on a large sign the expected time to the city center on each road. Then, we would have a model on which to base a decision.*

Perhaps this example illustrates the value of thinking about the parts of a model. One can argue that all we have done is use common sense. This is true. Indeed, much of modern technology and

*
If such a system were built, we would have to be careful. Otherwise, imagine that at 9 A.M. the signs say 60 minutes by the right road, 18 minutes by the left. Obviously, everybody goes left (except the stubborn driver who simply cannot stand technology). By 9:15 the signs read 30 minutes right, 80 minutes left. To avoid this, we have to calculate the times to take into account the probable behavior of motorists now reaching the "V."

Fig. 2–5.
An all-too-familiar sight
in an industrial area of a
major city. The reduction of
air pollution by legal action has
inevitable economic and social
consequences. New laws to curb
pollution must consider the
total effect on the city.
Otherwise, the changes
may be more harmful
than the pollution.
One help would be
to have a better idea
of which types of
air pollution are
most injurious to
humans. (The
Bettman
Archive)

mathematics is useful primarily because it guides us into good common sense.

Controlling Air Pollution

Reducing air pollution is a major goal of most city mayors today. The easy solution is obvious. We should first find out where the pollution comes from, then stop it at the source by appropriate laws.

If we attempt to do this, we find much of the pollution comes from cars, incinerators, and industrial stacks. We should ban cars, close incinerators, and stop industrial emissions. Unfortunately, we have used too simple a model. The mayor probably can do none of these.

Banning cars from the downtown area might be political suicide. As Henry Barnes, the late Commissioner of Traffic in New York, enjoyed saying, "Julius Caesar banned chariots in downtown Rome.* You know what happened to him." Even requiring taxis and city cars and buses to be electric means a tremendous expenditure of money.

Closing incinerators is no easier. If apartment houses must not burn rubbish, the city must provide for collection and disposal. Land-fill area in which to dump rubbish is rapidly disappearing, and several cities are already making plans to ship rubbish by train to distant locations. A cooperative effort between society and technology needs to be made to develop an efficient system for recycling trash.

Finally, forcing industry to clean stack emissions raises the cost of operating in the city. Many industries are already considering moving out of the big cities. An important job of the

*
Actually, population density in Rome was over 220,000 per square mile—far larger than today in any United States city, even though there were no high-rise buildings.

mayor is to increase the number of jobs in the city, particularly to take care of many relatively unskilled people moving in from other sections of the country. In many cities, the electric utility is a significant source of air pollution. Forcing the utility to clean its smoke raises the cost of electricity for both homes and industry. Again, industry is encouraged to move out.

Thus, decisions about air pollution require that we have a model which not only shows the sources, but also indicates the effects of legislation. The criterion must be not only clean air, but also a desirable total city. There must be jobs for people who want to work.

Perhaps this example illustrates the complexity of some of the decision problems we face today. The problem is even more complex, since it depends also on what we do in suburbia. If we install our traffic-information system of the first example, perhaps we will encourage more people to drive into the central city. In such problems, we have to determine a suitable model, even if this step requires great work. We cannot make decisions without consideration of the many different effects.

3 | ALGORITHMS

You and one opponent are playing the following game. You are seated across a table from each other. A pile of 11 matches is on the table. You and your opponent alternate turns. At his turn, the player may pick up 1, 2, or 3 matches (never 0 or more than 3). The winner is the one who forces his opponent to take the *last* match.

For example, one playing of the game might see the following moves:

You take 3	8 left
Opponent takes 1	7 left
You take 2	5 left
Opponent takes 1	4 left
You take 3	1 left
Opponent takes 1	0 left

Opponent moved last, you win.

Is there a strategy you can use which will ensure that you always win? In other words, are there rules you should follow? Such a rule might be: Always take 1 if there is an even number left, 2 if there is an odd number left, never 3. (This is not the correct rule.)

Such a set of rules is called an *algorithm*. Before we look at any more decision problems, we need to define this new term, algorithm.* *An algorithm is a sequence or a list of steps that can be taken to solve all problems of a particular type.*

*
The term algorithm is derived from an Uzbek (ŏoz'bek) mathematician, al-Khwarizmi, who developed such sets of rules in the ninth century. (The Uzbek Soviet Socialist Republic is a portion of the Soviet Union north of Afghanistan.)

We solve most decision problems by using algorithms. Consequently, in this section we want to become familiar with various algorithms and understand how they are used.

An Algorithm for Determining Averages

A very simple algorithm is that for finding the average of a set of measurements. The algorithm is described as follows:

1. Add the given measurements.
2. Count the given measurements.
3. Divide the answer to (1) by the answer to (2).

For example, if the given measurements are -2, 8, and 21, we find the average in three steps:

1. The sum is $-2 + 8 + 21 = 27$.
2. The number of measurements is 3.
3. The average is $\frac{27}{3}$ or 9.

The algorithm works for all possible sets of measurements.

An Algorithm for Number Selection

As we shall see, not all algorithms are so simple. It is not unusual for an algorithm to have hundreds, or even thousands, of steps. Regardless of size, an algorithm is a step-by-step procedure to solve a problem. Thus, an algorithm is just what we need to give instructions to a machine or computer. A computer can only do exactly as instructed by man. Hence, for a computer to solve a problem, we must first put into the computer the algorithm for that class of problems.

People also think in terms of algorithms. For example, the problem is to find the largest of the following positive integers:

13, 6, 47, 21, 5, 3, 7, 33

We first scan the list and eliminate all one-digit numbers, leaving

13, 47, 21, 33

Next we eliminate all two-digit numbers. There is nothing left. Then we back up one step and consider the two-digit numbers. The answer is among these:

13, 47, 21, 33

Now we can proceed in several ways. We might compare the first to the second. 47 is larger than 13, so we throw out 13. This leaves

47, 21, 33

Comparing the first two and throwing out 21 leaves

47, 33

Finally, the answer is 47.

Having worked this example, we can now write an algorithm for solving all such problems described by:

Find the largest of a given set of positive integers.

The algorithm must work for *any* problem of this type. For example, the algorithm must work for: 13, 8, 12, 13.

An interesting feature of the preceding problem is that the algorithm can have several different forms. For example, we might just compare the first two numbers and throw out the smaller. Then we continue this process until only the answer is left.

It is really not surprising that there may be several different algorithms for solving the same problem. Different people certainly solve the same problem in different ways. Some people are very clever at finding algorithms which give an answer quickly. Other people tend to use familiar approaches which may require a much longer time.

An Algorithm for a Decision Problem

The first two examples of algorithms were for number or arithmetic problems. Algorithms are also useful for solving problems which are not easily put into mathematical terms.

Consider the following problem: A town precinct covers a grid of streets (Fig. 2–6). A police car starts at A, where the police station is located. We wish to route this car so that all streets will be patrolled once. The patrol is to be completed in a minimum time so that the next trip can start as soon as possible.

We can cover the precinct in minimum time if we can find a route which covers each block once only. If this is not possible, the car should retrace as few blocks as necessary (or those blocks in which travel time is least). The problem is then: By inspecting Fig. 2–6, find a route which covers each block once only.

Exactly the same problem faces the mailman as he chooses his route, particularly when he travels by car and all mailboxes are at the curb. The newsboy and milkman have the same problem. Before considering a decision problem of such complexity, however, we will look at a much simpler version.

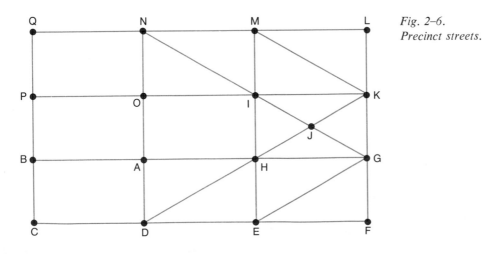

Fig. 2–6.
Precinct streets.

Early in the eighteenth century, the city of Königsberg (now part of the Soviet Union) included seven bridges over the Pregel River (as shown in Fig. 2–7a). A favorite pastime on a pleasant Sunday afternoon was to start at any point in the city and to see if one could walk in such a way as to *cross each bridge just once.* One could start at any point and end at any point.

In 1736, Leonhard Euler* spoiled this Sunday afternoon pastime of the Königsbergers by answering the question mathematically with an algorithm which solved the problem for any number of bridges. In order to understand the solution, we consider the map of Fig. 2–7a. There is an island C, two sides B and D of the main river, and a region A between the two sections of the river after it divides.

The interesting parts of the map are the bridges. Therefore, we simplify the map by showing only the seven bridges. Each bridge connects A to B, or A to C, or B to C, and so on. Let us make A, B, C, and D points and show only the bridges connecting these regions or points. This is done in Fig. 2–7b.

Each line in Fig. 2–7b represents a bridge. Each point, called a *vertex,* represents a land region. The graph has four vertices and seven lines.

Now suppose we can travel over each bridge once and only once. If the total route starts from A, we might go first to B. We would enter B on one line and leave on another line. Thus, *each time we pass through a vertex, there must be exactly two lines connected to that vertex.*

After the total path is drawn, we can say: Every vertex has an even number of lines connected to it. The only possible exceptions are the vertex from which we start and the vertex at which we end (e.g., if we start at A and go to B and so on, we leave A along one line; thereafter, we may pass through A again and two lines to A are added).†

Thus, a closed path covering all bridges is possible only if:

1. Every vertex has an even number of lines (then we start and end at the same point), or
2. Exactly two vertices have an odd number of lines (then we start at one of these and end at the other).

Inspection of Fig. 2–7b reveals that the vertices have the following numbers of lines:

$$A\ 3 \quad C\ 5$$
$$B\ 3 \quad D\ 3$$

All four numbers are odd. Hence there is no hope of walking over all seven Königsberg bridges without retracing any path. It makes no difference where we start and end or what route we follow.

Fig. 2–7.
Königsberg-bridge problem.

a) Seven bridges of Königsberg.

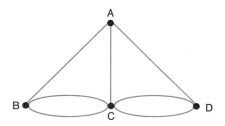

b) Graph of part (a).

*
The most famous of Swiss mathematicians who published an enormous number of articles on all phases of mathematics and physics.

†
If we start and end anywhere between vertices, every vertex must possess an even number of lines.

If one additional bridge were built (or we were permitted to swim across one river), the graph would be changed by adding one link or path between two regions. If the bridge were built from A to C, the graph would be as shown in Fig. 2–8. The problem can now be solved (Euler also showed this part of the solution). Now only vertices B and D are odd (as shown by the numbers in the figure). Hence, we can start from B, cross all bridges, and end at D.

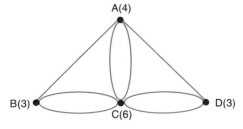

Fig. 2–8.
Königsberg with one additional bridge.

Can we find an algorithm which works in all cases? Euler's solution is as follows:

1. Follow any path from B to D (e.g., BAD).
2. Redraw the graph and omit the path already traced (Fig. 2–9).
3. From any vertex along this omitted (BAD) path, trace a closed loop (e.g., BCB).
4. Redraw the graph and omit this loop which has just been traced (Fig. 2–10).
5. Continue this process until all of the original graph has been covered (e.g., we might next trace the double loop ACDCA).
6. A solution now consists of the following steps. We start at B. We first follow all loops out of B (in our example, BCB only in Step 3). We then proceed to the next vertex on the main path. We continue this process until the entire diagram is covered and we have reached the terminal point. At each stage, we can redraw the graph in order to keep track of the positions which have been traveled. In other words, our solution says we should travel in the sequence:

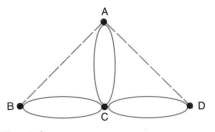

Fig. 2–9.
Graph redrawn with original path omitted.

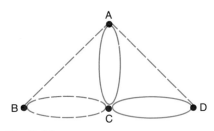

Fig. 2–10.
Graph redrawn with first loop omitted.

> B C B   First loop
> B to A
> A C D C A   Second loop
> A to D

We might show this by indicating the vertices encountered in sequence

> B C B A C D C A D

in order to cross all eight bridges with no retracing of steps.

Thus, for this example we have derived two algorithms. The first decides whether a solution is possible and also tells us where we can start and end. The second determines the route to be followed.

Routing Police Cars

We now try to apply the algorithm just discussed to the problem of routing police cars at the beginning of this section. The

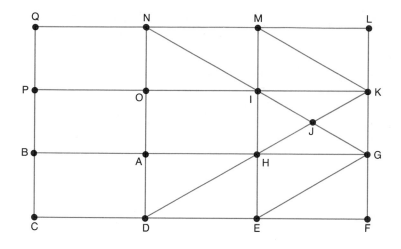

Fig. 2–11. Precinct streets.

precinct includes the grid of streets (Fig. 2–11). A police car starts at A, where the police station is located. We wish to route this car so that all streets will be patrolled once, preferably in a minimum time (so that the second patrolling can be started as soon as possible). We can cover the precinct in minimum time if we can find a route which covers each block once only. If this is not possible (as in this problem), the route must retrace as few blocks as possible (or select those blocks in which travel time is minimum). In order to find a desirable route for the car, we can apply the method of the preceding bridge problem. For this case, every intersection (marked by a letter in Fig. 2–11) is a vertex.

Ideally, we should like to start from A, travel all streets once, and end at A. Inspection of the figure reveals that this is impossible because the following vertices have an odd number of lines:

$$B \ 3 \quad K \ 5$$
$$G \ 5 \quad P \ 3$$

In order to permit a path from A back to A and through all streets, we must add paths which change these odd numbers to even (and leave the others even). In other words, we can add an extra path from P to B and from G to K. The car will traverse the P-B and K-G blocks twice during the patrol. This change is shown in Fig. 2–12.

Now the route-selection algorithm of the bridge example can be used to find an appropriate route for the patrol car. Two comments are important in this particular example:

1. In a more complex problem, there may be numerous widely separated vertices with an odd number of lines. In such a case, there are many different ways to add extra lines (retracing of certain streets) to make all vertices even. If

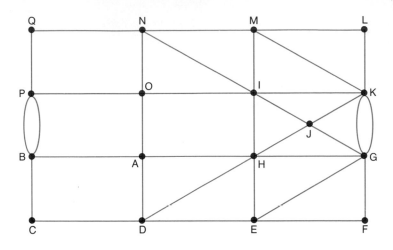

Fig. 2–12.
Street map
with added paths
from P *to* B, K *to* G
(corresponding to
two traversals
of these two
blocks).

we wish a patrol requiring minimum time, we must evaluate the extra time required for each possible retracing.

2. We normally wish to vary the patrol pattern of the police car, so that the patrol car does not pass a particular point at regular intervals of time. We can do this if we find different solutions from the route-selection algorithm.

An Algorithm for Games

As discussed at the beginning of Section 3, algorithms can also be used for games. Consider the following game which is more complex than our 11-match game. There are 27 matches on a table and two players, A and B. The players *alternately* pick up and keep 1, 2, 3, or 4 matches. Each player knows how many matches the other has at all times. The winner is the player who has an even number of matches at the end of the game.

In a game such as this, an algorithm for either player A or B consists of a set of strategies for the player to follow to make sure that he will always win or, if this is not possible, that he will maximize his chances of winning. In this particular game, if A goes first, he can always win if he moves or plays according to the following algorithm:

1. On the first move, A picks up 2 matches.
2. At each move thereafter, A proceeds as follows:
   a) If B has an even number: Divide number of matches still on the table by 6 and find the remainder of this division. Take one less than the remainder.
   b) If B has an odd number: Take one more than the remainder unless the remainder is 4, in which case take four (as soon as there are 1 or 3 matches on the table and B has an odd number, A should take them all).

This example of an algorithm demonstrates that there may be alternate paths which we have to follow to find a solution. In this case, the strategy to be used by player A depends on whether player B has an odd or an even number of matches. However, if player A plays according to the algorithm, he will win regardless of the path followed.

This example also illustrates that there may be several different forms in which the algorithm can be stated. In this case we might simply list all possible ways the game may develop (the number of matches remaining on the table after each move is shown in a circle).

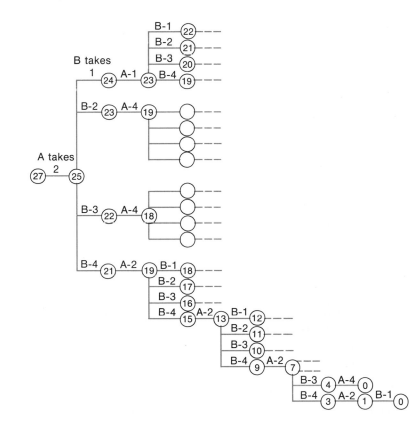

Chart 2–1.

Chart 2–1 is incomplete as drawn; if it were complete, it would indicate all possible ways in which the game might go if A always plays according to the algorithm.

In this section, we have considered several examples of algorithms. The decision-making problems which we look at in the remainder of this chapter and in the next chapter are solved in terms of algorithms. An algorithm is a set of rules for going from the statement of the problem (the model, constraints, and criteria) to the best solution.

4 | CRITERIA

In Section 1, the four elements of the decision-making problem are: model, criteria, constraints, and optimization. In the preceding section, we have discussed the optimization part, particularly when it is possible to find an algorithm or logical, complete procedure to a solution. In this section, we consider the *criteria*, or what it is that we want to optimize.

Probably the most significant decision made by most men throughout their lives is the selection of a wife. In this case, if one were rational and logical, a man might try to look at the problem in the following way:

1. Model: description of the characteristics of different girls. (appearance, beauty, intelligence, personality, character, aesthetic preferences, sense of values, attitude toward marriage, etc.)
2. Criteria: the relative values given to each of the above characteristics by the young man.
3. Constraints: the limited number of girls who would accept a proposal. If this number is precisely one, the problem is easy.
4. Optimization: selection of the girl to maximize the criteria within the constraints.

This particular problem perhaps illustrates the fact that most decision making in life is carried out intuitively and on the basis of poorly defined criteria. It is also clear that the two critical parts of an optimization problem are the model and the criteria— particularly the latter. While two young men might agree fairly well on the model (the different characteristics to be included and the relative ratings of particular girls), they would probably disagree strongly on the relative importance of these various characteristics. The criteria would vary greatly from one young man to the next. This is, of course, most fortunate; otherwise every man would want to marry the same type of girl.

In a simpler example (simpler because we can all agree on the criterion), we can illustrate another aspect of the criterion: the fact that we often have to change the criterion as the problem is being solved. In many cases, we discover that no solution can be found unless the criterion is simplified; in other problems, we can sharpen up the criterion as we move toward a solution. The latter situation is illustrated in the following example.

Figure 2–13 shows a model of a system of corridors in a building. There are three corridors: one from *a* to *b*, one from *a* to *c*, and one from *b* to *c*. These three corridors connect the points *a*, *b*, and *c*. During class changes in a school, the flow of people is from *a* to *b* in that corridor; hence, it is not convenient to try to walk from *b* toward *a*. We show this by placing an arrow from *a* toward *b*.

Similarly, the corridor from *b* to *c* is essentially one-way from *b* to *c*. The corridor from *a* to *c*, however, is wide, and we can move rapidly in either direction. This situation is represented by a double line connecting *a* and *c*, with an arrow in each direction. Thus, Fig. 2–13 represents the convenient pattern of travel throughout the system of corridors.

Now we turn to the problem we wish to solve. Our task is to station one or more men at the intersections *a*, *b*, and *c* in such a way that every intersection is covered by a man at most one "block" away. In other words, we might use three men, one at *a*, one at *b*, and one at *c*. Each man would then be responsible for monitoring his intersection.

Alternatively, we might place a man at *b*. He is able to cover both *b* and *c*, since he can travel to *c* rapidly in case he is needed there (there is a one-way path directly from *b* to *c*). Our man at *b* cannot, however, monitor intersection *a*. To reach *a*, he would have to travel two "blocks," from *b* to *c* and then from *c* to *a*. Hence, if we place one man at *b*, we must place another man at either *a* or *c* to cover intersection *a*.

The problem is where to locate the men in order to minimize the number required. The system of Fig. 2–13 is so simple that we can see that one man stationed at *a* could cover all three intersections (*a*, *b*, and *c*). If our only purpose is to solve this single problem, we are now finished. It is not difficult, however, to imagine a much more complicated pattern of corridors in which the answer is not obvious. In the hope of being able to solve problems of any complexity, we consider this simple example and look for an algorithm which works for all problems of this type.

The Corridor Algorithm

We must station our men to cover, or monitor, each intersection. We first consider intersection *a*. This can be covered by a man stationed at *a* or at *c*. We represent coverage of *a* by the expression

To cover *Man at*
    a $(a + c)$

Here the expression $(a + c)$ means a man at either *a* or *c*; the $+$ sign means the same as "or" in ordinary English. In other words, $a + b + c$ would mean *a or b or c*.*

Similarly, intersection *b* can be covered by a man at *a* or at *b*, represented by

To cover *Man at*
    b $(a + b)$

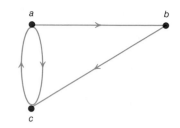

Fig. 2–13. A set of corridors.

*George Boole, 1816–1864, an English mathematician, introduced this idea when he showed how to represent statements in logic by algebra. Boolean algebra is routinely used today in analyzing and designing switching circuits such as those in a computer.

Now to cover both intersection a and intersection b, we can represent the stationing of men by the product

$$(a + c) \cdot (a + b)$$

Coverage   Coverage
of a of b

We are using the product to represent coverage of both intersections. In the algebra Boole developed, multiplication is read as *and*. Thus, $a \cdot b$ means a and b.

The above expression thus means we can have coverage of both intersections a and b by having a man at a or c, and a man at a or b.

Once this algebraic expression is written, we can work with it just as in ordinary algebra. If we multiply out, we have

$$(a + c) \cdot (a + b) = a^2 + ab + ca + cb$$

What does each term on the right mean?

The first term, a^2, means that a man at a covers intersection a (from $a + c$), *and* a man at a can cover intersection b (from $a + b$). The second term, ab, means the man at a covers intersection a (from $a + c$), *and* a man at b covers intersection b (from $a + b$). Similarly, ca means a man at c covers intersection a (from $a + c$), while a man at a covers intersection b (from $a + b$); cb means the man at c covers a, at b covers b. In other words, the product

$$(a + c) \cdot (a + b)$$

represents all possible ways of covering intersections a and b.

To cover all three intersections, we have

$$(a + c) \cdot (a + b) \cdot (a + b + c)$$

Coverage   Coverage   Coverage
of a of b of c

If we multiply out this product, we find one term, a^3, involving only one letter. In other words, there is just one location a at which we can place a man who can cover all three of the intersections, as we saw by simple inspection of the figure.

The problem of Fig. 2–13 is too simple to show the power of this method. Let's turn to Fig. 2–14, where we have six intersections ($a, b, c, d, e,$ and f). This might again be a problem of stationing monitors in school corridors. Exactly the same problem arises if we are stationing policemen on street corners.

In our example, we want to place men as needed to cover the important intersections $a, b, c,$ and d. Each of these four intersections must be within one block of a man. Intersections e and f are less important. (We might have found very little trouble at these

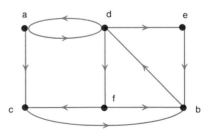

Fig. 2–14. Corridor map.

two locations. Consequently, it is all right if a man has to travel two or more blocks to reach *e* or *f*.)

The easiest solution is to put a man at *a*, another at *b*, one at *c*, and one at *d*. We want to *provide the required coverage with a minimum number of men.*

The statement of the problem is now complete. We must cover each of the four intersections, *a*, *b*, *c*, and *d*. Looking at Fig. 2–14 we see that intersection *a* can be covered by a man stationed at intersection *a* or *d*, intersection *b* can be covered by a man stationed at *b* or *c* or *e* or *f*, and so on. When our short-hand description of the problem covers the four essential intersections, we have the following expressions for the stationing of men:

$$(a + d) \quad (b + c + e + f) \quad (a + c + f) \quad (a + b + d)$$

To cover a To cover b To cover c To cover d

By multiplying this out, we obtain 72 terms ($2 \times 4 \times 3 \times 3$), each of which represents a suitable location of men to provide the required coverage. A few of the terms are identical, so there are slightly less than 72 different possibilities.

In this horribly complicated product, typical terms are,

$$dcab \quad a^2bc \quad a^2b^2$$

The term *dcab* means a man at *d*, one at *c*, one at *a*, and one at *b*. The term a^2bc means men at *a*, *b*, and *c* with the man at *a* covering two intersections. Since there are terms with only two letters, we can provide the required coverage with two men only; we can forget about all terms in the product with more than two letters.

The product can then be written*

$$(a + d)(b + c + e + f)(a + c + f)(a + b + d) =$$
$$a^3b + a^3c + a^3f + a^3e + a^2b^2 + c^2d^2 + a^2c^2 + a^2f^2 + d^2f^2$$
$$+ \text{terms with more than two letters.}$$

We now have nine different ways to station two men to provide the coverage required. The meaning of the terms is illustrated by the following two examples:

a^3b The man at *a* covers three intersections, that at *b* covers one;

c^2d^2 The man at *c* covers two intersections, that at *d* covers two.

The existence of nine solutions is an unexpected delight. In practice, however, we must choose one of these. Apparently we did not ask for enough in the original choice of a criterion. Instead of just asking for the minimum number of men, we can add additional criteria (i.e., we can sharpen our original criterion).

* These terms can be written without the tedium of writing out all 72 terms. Each product term contains one element from each factor; we merely look for combinations involving two letters only. For example, *a* appears in the first, third, and fourth term. Hence a^3 goes with any element in the second term. At this point, some textbook authors would leave the listing of all 72 factors as "an exercise for the student." What we like about applied science is that it is fun; hence we try to find those terms we need merely by inspection of the product.

We might require that each man be responsible for only two intersections. In other words, we ask for as even a division of the work load as possible. In our example, this added criterion rules out the terms $a^3(b + c + f + e)$ and leaves us with only the five possible solutions:

$$a^2b^2 + c^2d^2 + a^2c^2 + a^2f^2 + d^2f^2$$

Even with five solutions, we are unusually wealthy and we can add another criterion. We ask that each man be responsible for the intersection at which he is stationed plus one other (in order to minimize his travel time). This condition rules out a^2f^2 and d^2f^2, since in both cases the man at f is responsible for intersections b and c. We then have three possible solutions:

$$a^2b^2 + c^2d^2 + a^2c^2$$

Finally, we might select one of these three merely on the basis of which location is most important or possibly the personal desires of the men involved. The men we are stationing might prefer not to be placed at a because it is in a drafty location, so we would choose c and d.

General Comments

The interesting part of this problem is the way the final criterion was gradually worked out. We started with the simple criterion of minimizing the number of men. We found that we could add additional parts to the criterion: evening the work load, minimizing required travel, and finally catering to personal preferences.

In most optimization problems the initial criterion should be as simple as possible. Once solutions are found, we can add more criteria to the problem.

The question of what makes a satisfactory criterion underlies a large part of decision making. The continuing debate in the United States on the need for new weapon systems (for example, anti-missile missiles) stems from different criteria. Many of the people who want to build anti-missile missiles are worried about the defense of the country. They feel that, if we are strong defensively, we can prevent war. Opponents, on the other hand, argue that our economic and social needs must be faced. We cannot afford to meet these needs and engage in an armament race too. The difficulty of decision arises because of the constraints imposed by the need to limit government spending to control the national economic picture. When radically different criteria are used in decision making, it is not surprising that different decisions are reached.

In Section 4 we emphasize the importance of the criteria in decision problems. Of our four elements—model, constraints, criteria, optimization—the model represents the system with which

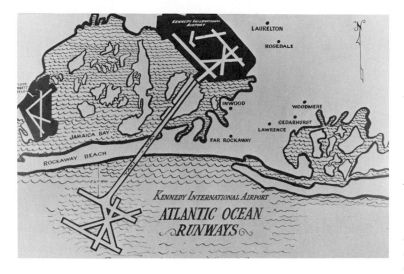

Fig. 2–15.
The choice
of an airport
location is one of
the most difficult decision
problems faced today. No
homeowner wants a jet airport
near his home. The noise is
continually annoying and may actually
lead to physical and mental illness. An
agency planning a new airport wants to
place it as close to the city center
as possible.

The figure illustrates one possible
solution for airports near an ocean
or lake. The runways for landing
and take-off are located well
out in the water, with a
taxi-way to the terminals.
The only significant
problem with such a
solution is the cost.
("Toward a Quieter
City," report of
the Mayor's
Task Force on
Noise Control,
New York
City,
1970.)

we are working, the constraints represent the limitations imposed on the system, the criteria represent the objectives of the system, and the optimization is the best solution to the problem. In applied science, as well as in personal decisions, the criterion is that part of the problem which is most difficult to describe precisely.

5 | OPTIMIZATION WITH FEW ALTERNATIVES

When optimization involves only one choice from a few alternatives, the selection can often be made by a simple comparison. In this section we consider two examples of such an approach. The first involves only two alternatives and the second requires a selection from among a small set of alternatives.

Choice from Two Alternatives

If there are only two feasible designs or plans, selection of the better of the two is usually simple if the people in charge of the design can agree on the criterion (that is, the basis on which the two possibilities are to be compared). For example, if there are just two possible choices for the location of a new airport, the only differences between the two might be as shown in this summary:

Location	Total cost	Time from center of town
East of town	$3,000,000	45 minutes
West of town	$5,000,000	25 minutes

The choice between the two locations depends on the importance of each of the factors. If we decide that the location nearer town will result in increased use of the airport and hence a much greater

(*Max Gschwind for* Fortune *magazine*)

*Fig. 2–16a.
The most
common method
of building in
water is to use land
fill. When the water
is not too deep, dirt or
solid material is used to
make an artificial island. Stone
is used around the sides
to prevent washing
away the fill. The lower
figure shows a
16-acre island
in Lake Huron
built for an
electric
generating
station.
(Photo
The Detroit
Edison Company)*

(*Max Gschwind for* Fortune *magazine*)

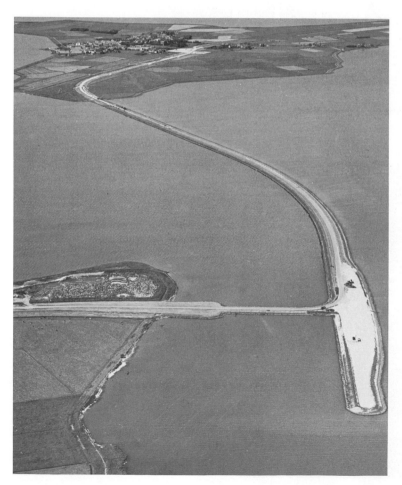

Fig. 2–16b.
Dry land can
be created by building
dikes to surround the area
to be made dry. Seepage through
the dikes is pumped out. The lower
picture is a scene from the
Netherlands, where more
than 1,600,000 acres
have been claimed
from the sea. (Photo
Netherlands
Information
Service)

(*Max Gschwind for* Fortune *magazine*)

Fig. 2–16c.
Piles can be
used to support
buildings. Much of Venice rests
on such columns. The deepest
structure is an oil platform
in 340 feet of water in the
Gulf of Mexico. Shown
here is a sulphur mine
seven miles off the
Louisiana shore.
(*Photo Freeport*
Minerals Co.)

(*Max Gschwind for* Fortune *magazine*)

Fig. 2–16d.
Floating structures
support airports, oil rigs,
and housing projects (as shown in
the model). The structure
must be even stronger
than a ship, since it
must stay steady
during waves. (Photo
R. Buckminster
Fuller)

tax and revenue return to the town, we may agree on the west location.

This class of problems in which there are only two alternatives is illustrated by a simple route-planning problem.

Let us look at Fig. 2–17. Bill is late for a date with Jean. He is anxious to drive from his house to Jean's house in the minimum time. By the rural road, which is six miles long, he can drive directly to Jean's house averaging 30 miles per hour. Driving over a winding feeder road, he can reach a parkway on which he is permitted to drive at the rate of 60 miles per hour. He then takes another feeder road to Jean's house. He can drive at 30 mph on the feeder roads which are each one mile long. The parkway section is six miles long, so the total route is 6 miles + 2 miles = 8 miles, as compared to the six miles for the rural road. Which road should Bill take?

The rural route can be driven in $\frac{6}{30} = \frac{1}{5}$ hour, or 12 minutes. On the other route, the two-mile trip along the feeder road can be completed in $\frac{2}{30} = \frac{1}{15}$ hour, or 4 minutes, and the 6 miles of parkway require $\frac{6}{60} = \frac{1}{10}$ hour, or 6 minutes: a total of 10 minutes. Thus the longer route is two minutes shorter in time. Bill should take the parkway route because he is late and must use the faster route. If he were not late, he might choose the rural road because it requires less gasoline, or because he enjoys the scenery. Clearly the best route depends on the criterion that is used.

This problem may seem rather unexciting: one can hardly care very much about a time difference of a few minutes. If we were dispatching fire equipment from a fire station to a fire, however, the saving of a few minutes by optimum routing could be of critical importance.

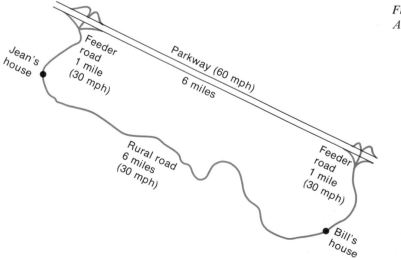

Fig. 2–17.
A route-planning program.

Choice from a Few Alternatives

If a small number of designs or plans is available, it is possible to study each plan carefully and to select that plan which is best. This is sometimes referred to as the "brute force" approach. To illustrate this method, we consider a very common *replacement problem:* When should one "trade in" his car to keep the average yearly cost as small as possible?

The problem is simpler if we consider trade-ins only after 1, 2, 3, 4, 5, . . . years of use. We must estimate the cost of operation of the car for each year of use. This cost must include the cost of fuel, oil, tires, batteries, maintenance, insurance, registration fees, taxes, and repairs. The cost of operation depends on the number of miles the car is to be driven each year, since this use determines depreciation, repairs, and so forth.

In Table 2–1 we list the data (columns 2 and 3) and the computations (columns 4 to 7) needed to answer our question. In column 1 we list the *age* of the car, and in column 2 the estimate of the *sale value* of the car. The car is assumed to depreciate $600 during the first year. The actual value depends on the make of the car, the market for second-hand cars, the mileage driven, and the quality of the care given to the car.

Column 3 lists estimates of the *operating expense* for each year. The operating costs listed are low and imply that the car is driven only about 10,000 miles per year. Operating costs do increase slightly as the years go by because of more repairs and poorer performance.

Column 4 is the *depreciation.* This is just the original cost ($3000) minus the sale value in column 2. Column 5 is the *cumulative operating cost*, which is the total of the operating costs in

Table 2–1.

Year (1)	Sale value (2)	Oper. cost (3)	Depreciation (4)	Cumu. oper. cost (5)	Total cost (6)	Av. cost per year (7)	
0	$3000	—	—	—	—	—	
1	2400	$ 800	$ 600	$ 800	$ 1400	$1400	
2	1920	850	1080	1650	2730	1365	
3	1540	900	1460	2550	4010	1337	
4	1230	950	1770	3500	5270	1318	
5	980	1000	2020	4500	6520	1304	
6	740	1050	2260	5550	7810	1302	← Sell
7	520	1100	2480	6650	9130	1304	
8	320	1150	2680	7800	10480	1310	
9	150	1200	2850	9000	11850	1317	
10	50	1250	2950	10250	13200	1320	

column 3 up to and including the present year. Column 6 is the *total cost* which is the sum of the depreciation and the cumulative operating cost (sum of columns 4 and 5).

The last column is the *average cost per year*. This is the total cost divided by the number of years. This column gives the cost per year if we trade in after 1, 2, 3, . . . years.

Thus, we want to find the minimum in the last column. The best operating cost per year is achieved if we trade in after six years. It is important to note that this minimum is very "flat"; it costs only $35 per year more (about 3% more) to trade in the car after the third year. If the reliability of operation or the status which comes from driving a late model car are considered, the car should be traded before the sixth year.

The decision is, of course, no better than the estimates that have been made of the sale value and the operating costs. The collection of the necessary data, upon which realistic estimates can be made, is the major difficulty in solving this problem. Furthermore, the solution of the problem does not take into account the probable future changes in the price of cars and in the cost of operation or the changing value of the dollar. As an extreme illustration of this latter factor, if severe runaway inflation should occur (as in Germany in the 1920's), money becomes worthless, costs rise at uncontrolled rates, and any economic analysis is useless days after it is made. Even a slight inflation, however, can change the best trade-in time.

The route-planning and trade-in examples described are very simple cases of optimization. In both cases, once a model is constructed (the data are collected), determining the optimum solution is straightforward. In the next section, we consider a more complex problem in which there are so many possibilities they cannot all be listed and compared one by one.

6 | DYNAMIC PROGRAMMING

In many problems, a large number of possible designs must be analyzed to find the optimum solution. A direct approach takes much too long. In such a case, we look for an algorithm which reduces the number of computations. We saw this approach in the Königsberg-bridge problem where tedious trial and error was eliminated by the use of Euler's algorithm.

Now let us consider an idea called *dynamic programming* which lets us find a simple way to solve a complex problem. Dynamic programming is an approach to those decision-making problems in which a sequence of decisions is required. Dynamic programming says: Once we know how to proceed in the best way from a particular point to the end, we follow this path regardless of how we reached the particular point. Consider the following example:

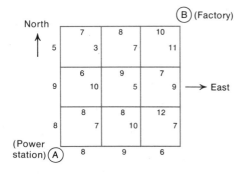

Fig. 2–18.
Costs per block
for new feeder line.

Fig. 2–19.
The street was built for three
lanes of traffic. Parked trucks
loading and unloading block both
curb lanes. The middle lane is very
close to blocked because of trucks
looking for a place to park. One block like
this obviously can result in a ten-minute
delay for an ambulance which happens
to pick this route. There just is no place
for the cars and trucks to go to leave
a clear path for the ambulance.

Angry citizens often urge a ban
on truck deliveries during the day.
This is really no solution. If trucks
have to deliver at night, labor
costs will increase. We can
expect higher costs of food,
clothing, and other necessities.
(Vincent J. Lopez for
Department of Water
Resources. "Toward a
Quieter City," report
of Mayor's Task Force
on Noise Control,
New York City, 1970.)

A power company, located at A in Fig. 2–18, plans to install a feeder line to a new factory, located at B. Streets have to be torn up and trenches dug for the new line. Some streets are paved with asphalt, others with concrete; the number of buried gas and telephone lines varies from street to street. On streets with heavy traffic, extra guards must be hired to direct traffic. The costs for the installation of the feeder line *in any particular block* are given by the numbers of Fig. 2–18. These range from a low of $3,000 to a high of $12,000. (Costs in the figure are in thousands of dollars.) How can we find which route from the power station at A to the factory at B will cost the least?*

This problem can be solved by calculating the cost for each possible route from A to B. We then can select the route with the minimum cost. This procedure is not particularly attractive, especially if we have a somewhat more complicated problem. For

*
In the text, the problem is to find the minimum-cost path from A to B in Fig. 2–18. The cost along each block is given.

The solution we find to this problem is equally useful for the problem of routing an ambulance from A to an accident at B in minimum time. Then the numbers in Fig. 2–18 represent the time to travel each block. In the same way we could talk about moving a fire truck from the station at A to a fire at B.

In a major city today, ambulances or fire trucks are often routed in a very illogical manner. The driver is given his destination and he selects the route he guesses will be fastest. He receives no information about traffic congestion along the various streets. It would be relatively inexpensive to measure traffic speeds along city streets at all times, feed these data to a computer, and have the computer use our algorithm to determine the best route. The decision would then be sent by radio to the driver.

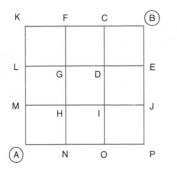

Fig. 2–20.
Grid of streets
with intersections labeled.

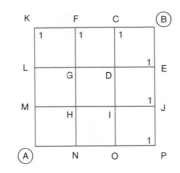

Fig. 2–21.
Number of routes
from intersections K, F, C,
E, J, P *indicated.*

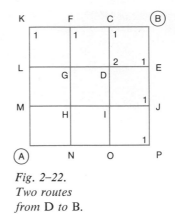

Fig. 2–22.
Two routes
from D *to* B.

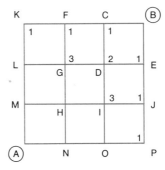

Fig. 2–23.
Three routes from G *or* I.

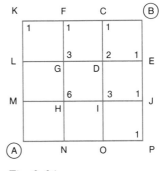

Fig. 2–24.
Six routes from H.

example, a grid eight blocks in each direction would have 12,870 different possible paths; calculating the cost for each would be a discouraging task.

Before we solve this problem by using a simple algorithm, let us digress briefly to see how the number of possible routes for any grid can be calculated. How did we obtain the number 12,870 of the last paragraph, or how many different routes are there for our 3 × 3 grid of Fig. 2–18?

Number of Alternate Routes

We assume that any route from A goes either to the north or to the east. In other words, no route moves south or west, and we always move *toward* the destination at B. There is a simple method for computing the number of possible routes.

The easiest method is to work backward from destination B. In Fig. 2–20 we have labeled each intersection with a letter. First consider intersections K, F, and C along the top. From each of these, there is only *one* route to B—east. Likewise, from each of E, J, and P there is only *one* route to B—north. Hence, we place a small one beside each of these intersections (Fig. 2–21).

Next we consider intersection D. From here we can go to E or to C; from either E or C there is only one route. Hence, from D there are two routes to B (Fig. 2–22).

If we continue to move away from B, we next consider G. Here if we go to D there are two routes, to F there is one. Hence, there are three routes from G to B. Similarly, we find three routes from I (Fig. 2–23). From H, there are three routes provided we go first to I, three if we go first to G—or six in all (Fig. 2–24).

In this manner, we find the total diagram of Fig. 2–25. There are twenty possible routes from A to B—twenty different costs to be evaluated and compared.* We have answered the question *by working from B back toward A.*

---

*
If the routes did not always move from A to B to the north or to the east, there would be any number of possible paths, since we could loop around a square block any number of times.

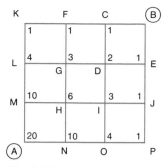

Fig. 2–25.
Number of routes
from each intersection
to B.

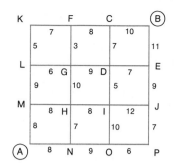

Fig. 2–26.
The original problem
with each intersection labeled
and costs per block shown.

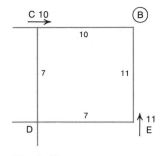

Fig. 2–27.
Best paths from C *and* E.

If the original problem had more blocks, we would find the following data:

No. blocks on a side	3	4	5	6	20
No. possible routes	20	70	252	924	137,846,528,820

The power company would be faced with a gigantic task if it had to figure out the costs for all possible routes from A to B for a grid twenty blocks in each direction. Consider the time it would take to figure out all the above possibilities. Some method of reducing the enormous number of possibilities is needed.

The Solution

The solution to the decision problem can be found if we work from the destination B backward, rather than starting at A and listing all paths. In other words, in Fig. 2–26, we first ask: If we are laying the pipeline and reach C, how do we proceed from here? We then repeat the question for E, which is also one block from B.

The answers to both of these questions are obvious. From C we must go east and the cost is 10; from E, north is the only allowable path and the cost is 11. We indicate these two "best solutions" on Fig. 2–27.

Next, how do we move if we are at intersection D? Here we can go north to C; from C we know east is the optimum. Hence north from D brings us to B with a total cost of $7 + 10$, or 17. East from D takes us to E at a cost of 7, then on to B at a cost of 11, a total cost of 18. Hence, if we are at D we should move north and the cost to B is 17 (Fig. 2–28).

We turn to the next set of intersections away from B: the intersections at F, G, H, I, and J. F and J are simple, since there is no choice (Fig. 2–29). G is calculated next. Going north yields a total cost of $3 + 18$ or 21; heading east yields $9 + 17$, or 26 (from

Fig. 2–28.
Best paths from D, E, C.

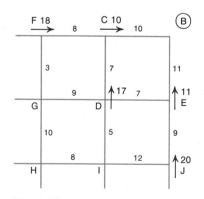

Fig. 2–29.
Optimum paths from F,
J, C, D, E.

D we already know the minimum cost is 17). Hence the desired path from G is north at a cost of 21.

In the same way the path from I is calculated as north at a cost of 22. Once I and G are fixed, H is calculated and we obtain the data of Fig. 2–30.

We next consider the intersections K, P, L, O, M, N, and A. This process finally leads to the minimum-cost values for every corner, shown in Fig. 2–31. Arrowheads at each corner show whether to travel north or east.

Thus, Fig. 2–31 is the solution of the original problem and gives the minimum-cost path from A to B. The total cost of this best path is $44,000.

The total grid has twenty possible paths from A to B. Instead of evaluating the twenty costs, one by one, we look at the problem as a set of binary (two-choice) decisions. There are nine of these in the total grid—at the intersections D, G, H, I, L, O, M, N, and A in Fig. 2–31. The solution of the problem requires the comparison of only nine pairs of numbers. These are, at each of these intersections, the cost if we go north or east.

To see how great this algorithm really is, let us look at the problem with twenty blocks each way. There are 137,846,528,820 possible paths. With the algorithm, we need compare only 20 × 20, or 400 pairs of numbers. If we can carry out four comparisons per minute, the algorithm solution takes less than two hours. If four path costs could be evaluated per minute (a rather high rate in a 20 × 20 problem), looking at all possible paths would take over 65,000 years of work. The algorithm turns an unsolvable problem into a solvable one.

Because a variety of optimization problems can be solved in the same way as the above example, it is useful to review the steps involved. We are faced with a problem in which there is a

Fig. 2–30.
Optimum paths from G, I, H added.

Fig. 2–31.
Minimum-cost values for the entire problem grid.

large number of alternatives. We try to find a way of looking at the problem as a *sequence of simple decisions*, rather than one complicated decision. We can find this sequence by working from the end B back toward the beginning A.

As an example, consider intersection H in Fig. 2–31. Before considering H, we determined the minimum-cost paths from G to B and from I to B (Fig. 2–32). If we are at H, we have only two choices, north to G or east to I. If we go north to G, we thereafter follow the already-determined, minimum-cost path to B (cost 21). Thus, from H the cost is 31 if we head north, 30 if we head east. We obviously should move east.

The key to success here is that, once we determine the minimum-cost path from G to B, it is the path to follow regardless of how we reach G. Recognizing this is what simplifies the decision at H.

There are many different kinds of optimization problems that can be solved with dynamic programming. Airlines take advantage of the wind direction on the different legs of the transatlantic routes to find minimum-time or minimum-fuel routes. Problems dealing with inventory control in warehouses, in programming of expansions of telephone switching centers, and in automatic control of industrial factories and chemical plants are other examples.

Fig. 2–32.
Decision at H,

7 | PROBLEMS WITHOUT SOLUTIONS

In a book, we of course look at problems which have solutions. Unfortunately, many real-life decision problems are not so simple. When we find the model, constraints, and criterion, we find that we have an impossible situation. In this section, we describe one important problem of this type.

Assume you are the owner of an apartment house in the center of a city. You are concerned about your property. The money you are receiving does not seem to be enough, and you must make a decision on what to do.

First, we describe the model. The building has five stories (no elevator) with eighty rooms. It is divided into twenty apartments, each with four rooms. There are 54 people living in the building. You bought the building in 1964 at a price of $80,000, which is its value. At that time, you invested $10,000 of your own money and the bank gave you two mortgages totalling $70,000. The annual interest on these mortgages is $10,500.

The city rent-control law allows you to charge $18 for each room each month. When all apartments are filled and everyone pays his rent, you receive $17,280 each year. To make repairs and keep the building in modest shape, you have to hire a full-time superintendent, someone part-time to collect the rents, and a lawyer part-time to help you meet all city laws. Also, you have to pay for repairs to the heating system and building. The cost of this maintenance comes to $15,000 per year.

We can summarize all the above in a table:

Investment	Income	Costs		Net loss each year
10,000	Rent 17,280	Mortgage payments	10,500	
		Maintenance	15,000	
		Total	$25,500	$8,220

Clearly, you are in trouble. You have invested $10,000 of your own money for the opportunity to lose $8,220 each year. A few more investments like this and you will be eligible for welfare payments.

Obviously you should raise the rent or cut down on maintenance costs. Here the constraints appear. The city rent-control law limits what you can do. If you want to raise the rents, you are told by your lawyer that the law has the following conditions:

> You are allowed to make 8% profit on your $80,000 building (or $6,400 each year). But you cannot count mortgage costs as part of your expenses. You cannot ask for a rent increase unless the rent is not enough to cover maintenance (yours is). Also, you must have lost money (not counting mortgage payments) two years in a row before you ask for higher rents.

Since your rent income is greater than maintenance costs, the law does not allow you to raise the rent.

Your other possibility to stop losing money is to cut maintenance costs. You hire the superintendent for only half time and try to cut repairs. The tenants complain, the city building inspector is called, he finds violations and recommends to the tenants that they pay only $1 per month until the repairs are made. Suddenly your rent income drops from $17,280 per year to $240 per year. Now you are really losing money!

You are hopelessly trapped. There just is no decision you can make to change your annual loss into a profit. You have only one possible decision: abandon the building.

Unfortunately, all the preceding numbers are not imaginary. They are typical 1970 data for the housing problem in New York City. They illustrate the economic basis of the problems which promise to create a national housing crisis in our cities in the 1970's.

As a result of this situation, 30,000 apartments are being abandoned each year in New York City and the number is growing. In the last three years of the 1960's more buildings were abandoned than could be torn down in twenty years of slum clearance. The city is filling up with abandoned buildings. With practically no new apartment construction, the 1970's are likely to see 25,000 families a year in New York City with no place to live.

Fig. 2–33.
A scene in East New York
is an example of the abandoning of
housing in the city.

They will have to live in armories or move out of the city, unless the model and constraints described can be changed in a major way.

The discussion above is presented in terms of the landlord. We might also have viewed this as a tenant. Faced with poor maintenance leading to inadequate heat, decay of the building, and dangers for his children, he has no recourse but to appeal to the city officials for help and to refuse to pay rent—steps which lead so often to abandoning of the building and his eviction. Likewise, we might take the view of the city government and consider the problems of writing a successful rent-control law.

The tendency to blame only the tenant, only the landlord, or only the city government clearly simplifies the problem. The model and constraints in this case are such that there is no optimization possible. To improve the situation, we must start anew with changes in the model and the constraints.

8 | FINAL COMMENT

Our goal in this chapter was to introduce the four elements of decision making. The three parts of the problem are:

MODEL   CONSTRAINTS   CRITERIA

Once we know these three parts, we can proceed to the

OPTIMIZATION

Several simple examples are used to show what each of these parts means.

From this chapter, two features of decision problems should also be clear:

First, decision problems are everywhere. The high school senior must decide whether to go on in school, what career to follow, what school to choose. The college senior again faces similar decisions. Marriage, choosing a job, picking a home, buying goods, deciding on a vacation—every major action we take involves a decision. Businesses and government also face decisions continually, often decisions which have a great effect on the future.

Obviously, as we think about a particular decision, we don't usually ask ourselves: What is the model? What are the constraints? What criteria should we use? Why then should we worry about the ideas in this chapter?

Some of the decisions we make are major factors in determining our future happiness, the relationship we have to other people, and our role in society. Because these decisions are of such importance, we must attempt to bring to them as much intelligence, wisdom, and logical thought as possible. Understanding the elements of a decision problem often helps in making a better decision.

Furthermore, government, business, and social organizations are more and more being run by men who try to use *logical* decision techniques. It is now common for a town to have a master plan for the future. Deciding on a plan requires a model, constraints, and a criterion. Even though estimates of the future are certain to be wrong, in order to make our decisions as logical as possible we must use a quantitative model. That is, we must describe our model in numbers. If we as voters and citizens are to understand government action, we must be able to see how the model, constraints, and criteria are measured.

Decision methods are often criticized because of this dependence on numbers. People ask, "Since the decision depends on predictions or guesses and since the numbers may be wrong, why is the decision any better than a wild guess or a hunch?" The basic idea of all modern system studies is that a model, even if imperfect, improves the decision. The more nearly correct the model, the better the decision. Hopefully, the decision will be good even if the model changes slightly.

The second interesting feature of decision problems is that there is no simple method for finding the answer. We looked at a few algorithms in this chapter. These work only for special kinds of problems. When we face an entirely new problem, we have to use our imagination and ingenuity to try to find an answer.

This situation is quite different from the usual problems considered in science and math texts. Often there we can learn a method for solving a whole category of problems (the quadratic formula in math). When we consider difficult problems from the real world, there often is no straightforward way to the answer. We have to try different attacks. We can only hope for success. Decision problems are challenging because they have this feature.

Questions for Study and Discussion

1. Test the steps of the algorithm which can be used to solve any problem of the Königsberg-bridge type. We can assume we start with a graph as a model for the system; the solution desired is either a statement that the problem is not solvable or a step-by-step procedure for finding a path which travels on every line once and only once.

2. The girl at the check-out counter in a supermarket has to make change. The usual method can be described by an algorithm, or by a set of algorithms for use in different situations. Work out algorithms for these cases:

   a) A dollar bill is given for a 22¢ purchase.

   b) A five-dollar bill is paid for goods costing $2.78.

3. What criteria would you use in selecting a piece of sporting equipment for yourself if money were not a constraint? Select any kind of equipment that you feel you are a good judge of, for example, skis, ski boots, baseball glove (for which position?), surfboard, tennis racket, 4-iron.

4. Suppose in the previous case that you are able to spend only 75% of the cost of the "best" item. Assume that you don't have to shop around, because the store carries every conceivable item in stock—if you can imagine such riches. What happens to your criteria? That is, do you relax all your criteria about equally, or do you eliminate or greatly weaken some requirements in order to retain others unchanged? (Be as specific as you can and see if you can produce "logical" reasons.)

5. In a small city in the Midwest, there is a three-mile stretch of one street on which the traffic lights are timed so that a motorist who drives at 30 mph can expect to be confronted by a red light only once or twice during his trip west on this street. Depending on the traffic density (number of cars per mile), the timing is changed so that 28 mph, 23 mph, or some other speed becomes the optimum speed. What constraints might affect such a system?

6. The following problem can be solved by the Königsberg-bridge algorithm. Explain what you would do to fit the algorithm to this problem. Problem: Draw a continuous line that passes through each edge of every rectangle in the diagram once and only once.

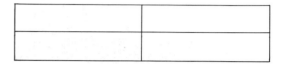

7. Apply the method you developed in Question 9 to find out whether the problem can be solved if the diagram changes to this:

8. Various performance criteria are appropriate in evaluating a record player. Discuss those which would be used:

   a) by a committee of students buying a record player for a school recreation room.

   b) by a restaurant owner planning to pipe background music into his various dining rooms.

   c) by a lover of contemporary "serious" music (the orchestra is likely to have a large percussion section).

9. Could a walker cross all the bridges of Fig. 2–8 (starting at B and crossing each bridge once and only once) by any other path than that given in the text? If so, describe one such path.

10. A small town has a grid of streets like the corridor map of Fig. 2–14. The town government decides to install street lights, but wishes to minimize cost. Intersections a, b, c, and d should be illuminated (more or less) in each case by a light not more than a block away, but e and f are considered less important. Give an expression which describes the possible placement of street lights. Do not expand the expression.

11. Figure 2–5 has a caption which says that the reduction of air pollution by legal action has in-

evitable economic and social consequences which may be more harmful than the pollution in the short run. Describe a pollution situation in which immediate solution would cause serious economic and social consequences. Describe a pollution situation in which the failure to solve it immediately would cause serious economic and social consequences.

12. You wish to run a wire from your record-player amplifier at A to a speaker at G in the room pictured. You must run the wire along the edges of the room. List all the possible ways to install the wires following the constraint given. What is the optimum path to use to insure that future shifts in furniture (other than the record player or speaker) would not require rewiring?

13. A young married man with one child, working for a "national company," is sent to a new town where he must buy or rent a house or apartment. What are likely to be (a) the constraints and (b) the criteria which will govern his decision?

14. Refer to Question 13. Would you expect either of these to change if the man were an executive officer of the company, aged about 55, with no children at home, but grandchildren not many miles away? If so, how?

Optimization

3

1 | INTRODUCTION

In this chapter, we will consider problems of three types. Type 1 is a *queueing* problem. Type 2 represents a group of problems described by the name *games*. Type 3 comes under the general heading of *linear programming*. While we will not go into the detail required to solve the problems described, we will look at methods of solution which can be used in such problems. All of these are decision problems. They have the elements described in the preceding chapter. All are solved by optimization. These decision problems are illustrated by two examples.

Problem 1. By 1980, 30 million people a year will be using Chicago's three airports. Three quarters of these people will go to the airport by car or taxi. How many lanes should we build for the main highways entering the airports in order to prevent excessive traffic jams except on very rare occasions? The answer turns out to be six. Once this answer is obtained, the airport designer next has to worry about parking places for all the cars.

Problem 2. We have limited total resources, too little to do everything we would like to do. How do we allocate these resources in an optimum decision? For example, a sausage-maker has a recipe for frankfurters which includes a number of ingredients for which the prices fluctuate widely from week to week. He wants a computer program to determine the proportions each week to maximize the profit. Of course, there are constraints to limit the variations of flavor so the customers do not object.

2 | QUEUEING PROBLEMS

A queue is a waiting line. An example of a queue is found at a theater box office just before a show starts. Some of the worst queues exist at a typical college when new freshmen arrive to register or at an army induction center.

For our discussion, we must recognize that queues are likely to form when customers arrive at a service facility. In order to be general, we talk about *customers* arriving. In actual examples, these customers may be people (arriving at a supermarket checkout counter) or machines (airplanes arriving at an airport or subway trains waiting to enter a station).

These customers are to be *serviced*. The service facility processes the customers, in many cases one at a time. Once the customer is serviced, he leaves the system.

A Supermarket Queueing Problem

A small supermarket is built with eight check-out counters. The manager is ready to open for business and is hiring people. He needs to know how many check-out girls to hire. How many of the eight counters should he plan to open?

If too few counters are open, long queues are likely to form. Customers who have to wait fifteen minutes to check out are not likely to come back to this store. On the other hand, if he opens too many counters, he will be paying the salaries of people who are working only a fraction of the time. Since the profit of a supermarket is not large under the best conditions, he cannot afford very many unproductive people.

This problem is interesting because it is complicated by changes which occur randomly, or without a definite pattern, during the day. In a supermarket, there are times when, for no apparent reason, the check-out system is very heavily loaded. At other times, hardly any customers are checking out. These changes are partly predictable (Friday evening is a busy time). They are also partly unpredictable. The strategy the manager chooses must take care of the unexpected load.

How does the manager make decisions? As we look at queueing problems in the next few sections, we will find some of the techniques which can be used to make the system adapt to occasional heavy loads. Three steps are obvious in a supermarket.

1. An express counter is set up for customers who do not require very long to service. This removes these customers from the queues at the slower service channels.
2. If the system becomes overloaded, we try to reduce the service time for each customer. In a supermarket, the manager rapidly switches some men from arranging merchandise on the floor to helping the check-out clerk by putting the items in bags. Once the queues are reduced, the men are returned to the other tasks.
3. Finally, the manager or other supervising personnel can man unused check-out counters.

As we discuss queueing, we will call these three strategies:

1. Changing the priority system (that is, which customers are served first).
2. Reducing the service time.
3. Increasing the number of service channels.

These are three basic techniques for keeping queue lengths small. In the supermarket case, they are easily carried out. In other problems, they may not be possible, especially if customers insist on being treated on a first-come, first-served basis.

"What did I say about leaving on a Friday afternoon!"

*Fig. 3–1.
A queueing
problem. (Drawing
by Stevenson;
© 1969
The New Yorker
Magazine,
Inc.)*

Examples of Other Queueing Problems

Queueing problems arise in such diverse situations as:

1. Tollbooths on a parkway, where long lines of cars may build up.
2. A barbershop or hairdresser operating without appointments. Even though the barber may be idle a significant part of the time, there are other times when a queue of three or four customers forms. The queue seldom becomes longer because customers turn away rather than wait.
3. An airport in which only one runway can be used for take-offs and landings. Because of traffic demands generated by men anxious to return home for the weekend, many major airports in the United States have very large queues of planes circling and of aircraft waiting to take-off every Friday afternoon around 5 P.M.
4. A telephone exchange. On Christmas and Mother's Day, long-distance toll lines are frequently saturated.*
5. A production facility dependent on a number (perhaps ten) of machines which must be maintained by two mechanics. Each machine breaks down occasionally and can be repaired by a mechanic. The amount of time required for servicing varies according to the nature of the difficulty. In this case, the queue consists of inoperative machines being repaired or waiting to be repaired. The customer for the service in this case is the machine.

*
The telephone-system example is particularly appropriate, since the basic work on queueing problems was originally developed for guiding the design of telephone systems. A. K. Erlang of the Copenhagen (Denmark) Telephone Company was a key worker as far back as 1907.

Importance of Random Behavior

Queueing problems are interesting because both the customer arrivals and the servicing times are random. They vary in an unpredictable way.

If the customers arrive at regular intervals and each is serviced in a specific time, the queueing problem is trivial. For example, in the case of a single service channel (Fig. 3–2), if customers arrive to be serviced every 5 minutes and servicing requires 4 minutes, we can state:

> The service channel is busy $\frac{4}{5}$ of the time. The maximum queue length is one (the person being serviced). The average queue length is $\frac{4}{5}$ because $\frac{4}{5}$ of the time it is one, and $\frac{1}{5}$ of the time zero.

There is really no queueing problem. If the service time were greater than 5 minutes, the queue would steadily grow larger. More and more people would be waiting longer and longer.

If random behavior is considered, the situation is quite different. For example, we can imagine that customers arrive completely at random with an average of one every 5 minutes. In other words, in 500 minutes we expect about 100 arrivals. These come, however, with equal probability in every short interval of time.

Furthermore, we can assume that *on the average* 4 minutes are required per customer. Any one customer may require more or less time. In other words, we assume that the servicing times vary at random in the same way that the times between customer arrivals vary.

If these two types of random behavior occur, the queueing problem is much more difficult. Now it is quite common for a group of customers to arrive very close to one another. The second, third, and fourth customers may arrive before the first is served. Indeed, if we study such situations, we find that over a

Fig. 3–2. Regular arrivals, single channel.

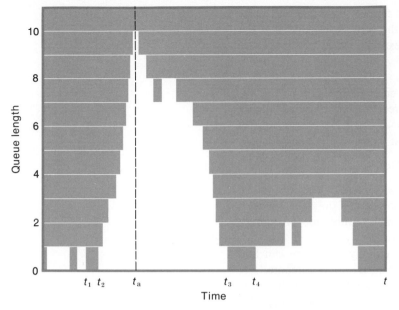

Fig. 3–3. Possible plot of queue length for random arrivals and service times.

long period of time, the average queue length is four. On the average, there are three people waiting and one being served, even though the service channel is still busy only $\frac{4}{5}$ of the time.

In this case, we can run an experiment and observe the way the queue length q changes with time. Figure 3–3 shows one such graph obtained from a computer study. From t_1 to t_2 and t_3 to t_4, there is no queue at all. The service channel sits idle. Right after t_2, customers arrive rapidly. By t_a, the queue length is ten.

Even though the service channel does not seem overloaded (it is idle $\frac{1}{5}$ of the time), the customers who arrived around t_a are obviously not well treated. The random feature creates the problem.

Because the randomness is so important, we can only study queueing problems experimentally with a few basic ideas of probability. These are presented in the next section. We will consider queueing again in Section 4.

3 | PROBABILITY

One of the striking features of the past twenty years is the growing awareness that many things can only be described in terms of probability. Even weather reports now often contain the statement, "The probability of rain tomorrow is 20%." The ideas of probability, which were first developed in the eighteenth century to analyze gambling games, are essential today for understanding many important systems, from traffic flow on highways and through tunnels to queues. In this section, we will discuss a few aspects of probability.

If we thoroughly shuffle a deck of 52 playing cards and draw one card at random from that deck, the probability that it is the 7 of spades is $\frac{1}{52}$. If we repeated the experiment 52,000 times, we would expect that the card selected would be the 7 of spades about 1000 times.

Can we describe this experiment more carefully? In the experiment, there are 52 *equally likely* results. The phrase "equally likely" is very important here: Each of the 52 cards is equally likely to be the one selected. We can list these equally likely results:

Spade A, 2, 3, . . . , 10, J, Q, K;   Heart A, 2, 3, . . . , K, . . .

Before the experiment is performed, we decide which results we will call favorable: In our example, we selected only the spade 7. Hence, of the 52 equally likely outcomes, only one is desired. Then the probability of a favorable result is $\frac{1}{52}$.

As long as the possible results are equally likely, we can find the probability of any group, rather than a single outcome. For example, the probability the card drawn is a 7 is $\frac{4}{52}$. (There are 4 sevens among the 52 possible outcomes.) The probability of a spade is $\frac{13}{52}$, of a face card (jack, queen, or king) is $\frac{12}{52}$.

Listing Equally Probable Outcomes

Thus, to determine the probability of a favorable result, we first list the equally likely outcomes. We must be careful that the items of the list are indeed equally likely. As an example, consider the problem of tossing two coins. There are now three possible results:

2 heads   1 head and 1 tail   2 tails

These three outcomes are not equally probable, however. If one coin is a penny P and the other a nickle N, the possible results are

P	N	Overall
H	H	2 heads
H	T	1 head and 1 tail
T	H	1 head and 1 tail
T	T	2 tails

Thus, the probability of two heads is $\frac{1}{4}$; the probability of a match is $\frac{2}{4}$, of a mismatch also $\frac{2}{4}$.

One way to calculate the probability is, therefore, to list all equally likely outcomes of the experiment, count these (call the number N), count the favorable results (call this n), and finally calculate the probability as

$$\frac{n}{N} \text{ or } \frac{\text{favorable results}}{\text{total, equally likely results}}$$

This method works only when the outcomes are equally probable.

Meaning of a Probability

What a certain probability means cannot be measured easily. It depends on the importance to you of the various outcomes of the experiment.

For example, you are entering the hospital for an operation. Statistics indicate the probability of full recovery is some number y, where y might be $\frac{8}{10}$, or $\frac{9}{10}$, or $\frac{999}{1000}$. Above what value of y do you stop worrying about the operation? Obviously there is some value above which you are safer in the hospital than going to work. Yet most of us would worry even if the probability of full recovery were one. The significance of probability depends strongly on the person's individual response to the various outcomes and on the value he places on each possible outcome.

As another example, a probability of 20% for rain tomorrow seems like an optimistic forecast if you are planning nothing unusual. It is worrisome if you have tickets for a World Series game and, if there is a postponement, you cannot go the following day.

Intuition and Probability

Intuition is not always much help in probability problems. Simple mathematical "illusions" are as common as optical illusions. In the most familiar example of such a mathematical illusion, an honest coin is tossed repeatedly with a sequence of heads and tails. As soon as four heads in a row come up, the tossing pauses while bets are placed on the single next toss. Most people would prefer to bet on a tail, and occasionally we can even find an "educated" individual who will give us favorable odds if we allow him to bet on a tail coming up next. Actually, the probability of a head coming up on the next toss is exactly $\frac{1}{2}$, since preceding tosses are irrelevant past history.

The examples of such breakdowns of common sense are numerous. As another example, consider a system in which a series of men check their hats as they enter a restaurant. Unfortunately, the checkroom attendant is hopelessly unintelligent and simply gives each man, as he leaves, a hat selected at random from her collection. What is the probability that no man will receive his own hat? In other words, how often does it happen that everyone receives the wrong hat?

We do not wish to solve this problem completely because the solution would take us into an area of mathematics which is not really important to the following chapters. The answer does depend, however, on the number of men who have checked hats. If only one man has checked his hat, there is no chance he will get the wrong hat. With two men, there is a 50% chance they both will; with three men, there is a 33% chance all three men will receive wrong hats; and we can continue enumerating possibilities.*

*
To obtain these results, it is easiest to count all the ways in which the hats might be distributed. We construct three rows: the first is the hat man A receives, the second B, and the third C. Then we call man A's hat a, man B's hat b, and man C's hat c. The hats can be distributed as follows:

To A—a a b b c c
To B—b c a c a b
To C—c b c a b a

The underline denotes the wrong hat. There are only two columns (fourth and fifth) with all entries underlined — two of the six possibilities (33%) in which *every* man has a wrong hat.

If the table is worked out for four men, it shows 24 different ways to return the hats. In 9 of them (38%) every hat ends on a wrong man.

As the number of men becomes very large, the answer approaches 37%.

The important feature of the problem is not the particular answer, but two aspects of the answer:

1. The probability of no one receiving his own hat is essentially the same if there are 8, 20, 800, or 4000 men. Once the number exceeds 8, the actual number is largely immaterial (37% of the time this experiment is tried, no one receives his own hat).
2. The probability is higher for an even number than for either adjacent odd number. In other words, the event is more likely to occur with four men than with either three or five, more likely with six than with five or seven.

Both these results are rather startling and counter to the intuition of most people.

Probability of Several Events

We need one more characteristic of probability. If two *independent* (unrelated) experiments are performed, the probability of success in both is the product of the individual probabilities.

As an example, in a football betting pool, you are asked to pick the winners of four different games played the same day. Each of the games is an even bet, that is each team is equally likely to win. In an actual pool, points are added to the score of the weaker team to make the bet even. The probability of your selecting a winner on each game is $\frac{1}{2}$; the chance of picking all four winners is

$$\frac{1}{2} \times \frac{1}{2} \times \frac{1}{2} \times \frac{1}{2} \text{ or } \frac{1}{16}$$

If you bet $1, you should receive $16 if you are successful. Typically, the man operating the pool pays off $10, so his profit averages $6 on each $16 bet. Actually, his profit may be even greater if a tie game results and all bettors lose.

We can also use this idea of independent experiments to calculate the probability of drawing the spade 7 from the shuffled deck of 52 cards. The suit drawn is independent of the number of the card. Hence, there is a probability of $\frac{1}{4}$ of drawing a spade, a probability of $\frac{1}{13}$ of drawing a 7. Success in both directions has a probability of

$$\frac{1}{4} \times \frac{1}{13} \text{ or } \frac{1}{52}$$

As another example, a high school senior applies for admission to four colleges. He knows the colleges act independently on admission decisions. He estimates his chances of admission at each college are

College A $\frac{1}{5}$ College C $\frac{1}{2}$
College B $\frac{1}{2}$ College D $\frac{5}{6}$

What chance does he have of being admitted to all four?

Since the four "experiments" are independent, the probability of success in all four is just the product of the individual probabilities:

$$\frac{1}{5} \times \frac{1}{2} \times \frac{1}{2} \times \frac{5}{6} \quad \text{or} \quad \frac{1}{24}$$

He has 1 chance in 24, which is a little over 4%, of receiving four acceptances.

What is the probability he will receive *no* acceptances? Now to "succeed" in this experiment, he needs to be rejected by all four colleges. At each, the rejection probability is*

| College A | $\frac{4}{5}$ | College C | $\frac{1}{2}$ |
| College B | $\frac{1}{2}$ | College D | $\frac{1}{6}$ |

The probability of four rejections is

$$\frac{4}{5} \times \frac{1}{2} \times \frac{1}{2} \times \frac{1}{6} \quad \text{or} \quad \frac{1}{30}$$

He has a probability of $\frac{29}{30}$ (about 97%) of receiving at least one acceptance. Again, what does this mean? Should he worry until he hears from the schools? Obviously, whether 97% success is favorable odds or not depends on the individual interpretation of the outcomes.

The college-admission example above illustrates a common class of problems: We are given the probabilities of success in various, independent experiments; we want the probability of success in *at least one* experiment. To solve this problem, we find the probabilities of failure in each, then the probability P of failure in all; finally $1 - P$ is the probability of succeeding in at least one experiment.

This same approach is used to solve a classic probability problem. In a room of N people, what is the probability that at least two have the same day of the year as birthdays? To solve this, we start around the room and ask each person in turn what his birthday is. The answer of the first man A is immaterial. The second man B has a probability

$$\frac{364}{365}$$

of having a different birthday; the third man C has a probability

$$\frac{363}{365}$$

of having a birthday different from *both* A and B. The probability that three men (A, B, C) have different birthdays is then

$$\frac{364}{365} \times \frac{363}{365}$$

*We assume at each college he is either admitted or rejected. Hence, the probability of rejection is one minus the probability of acceptance: if he has $\frac{5}{6}$ chance of being accepted, he has $\frac{1}{6}$ chance of being rejected.

Fig. 3–4.
Results
from a study
by freshman
engineering students
of a comparison of four
types of golf clubs (drivers).
For each club, the horizontal
axis shows the ball speed just
after it was hit. The vertical axis
indicates the number of times the
result was obtained. Thus, for Club 4A,
a speed of 180 feet/sec was measured
three times.

The data illustrate the probability
characteristic found in most
experiments, particularly when
human beings are part of the
system. The human golfer just
cannot hit the ball identically
every time. In such a case,
we can use the average
(*shown above*). *If we*
are golfers, we might
also be interested in
using the club with the
greatest consistency.
(Technology
Review of M.I.T.)

What is the probability that five people in a room all have different birthdays? What is the probability, with five people in the room, that two or more have the same birthdays?

If one continues this calculation for larger and larger groups of people, we find that with only 22 people in the room, the probability that they all have different birthdays is less than one half. Hence, if we have a room with 22 or more people, the chances are that at least two have the same birthday. We assume, of course, there are not identical twins in the room. Furthermore, we neglect the possibility of a February 29 birthday.

A similar technique can be used to solve the following problem. Although a complete analysis is much too tedious, the problem does illustrate the difficulty of guessing answers to probability problems. Often our guesses are not even close to the correct answer. Here is the problem.

A gum manufacturer puts a baseball card in with each package of gum. There are 100 different cards available, placed at random in the packages. Johnny decides to collect cards and asks his father, "Tell me, before I start buying gum, how many packs will I probably have to buy before I have the complete set of 100?"

Johnny's father answers, "If you do no trading with your friends, you can expect to have to buy _____." What number should be inserted in the blank? Make your own estimate before turning to the answer in the footnote at the end of this section. This is an interesting example, since scientists and engineers often guess far from the correct answer.

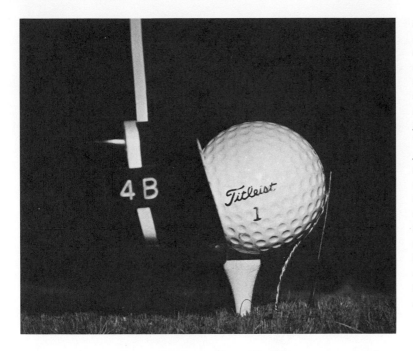

Fig. 3–5.
Picture of a golf club
and ball in contact as the
ball is being driven. As the ball
moves off the tee, it closes an
electrical switch at the right to
start a series of high-speed
photographs. The ball
velocity is determined
from successive pictures
taken a known time
apart. Of incidental
interest is the
astonishing extent
to which the ball
is compressed.
(B. L. Averbach,
M.I.T. 13—5082,
Cambridge,
Mass. 02139)

One final example illustrates the difficulties we can get into when we talk about situations governed by probabilities. Consider this problem. Three men stand 50 yards apart from one another in a field; each has a gun and bullets. We call the men:

S   for sharpshooter
M   for marksman
D   for duffer

Each shoots one bullet at a time in rotation: first D, then M, then S, then D again, and so forth. Each man can see whether the other two are still alive and standing, each can hear the others' shots, but they cannot see who the others are aiming at as they shoot.

They have quite different abilities. S always hits what he aims at (the probability of a hit is 1), M hits 80% of the time, and D hits 60% of the time. If D shoots at S and misses, S will hear the shot and know D has shot at him.

D shoots first. Where should he aim?

Then M shoots if he is still alive. Where should he aim?

Then S shoots if he is alive. Where should he aim?

In other words, you should take the part of each of the men in turn and decide on the best strategy.

The answer for D is easy. He certainly does not want to shoot at M; if he should hit M, S would certainly shoot him, and the three-way duel would end with S the sole survivor. Likewise, D probably should not shoot at S. If he misses and M does not hit S, S will be apt to aim at D in revenge. If D happens to shoot S,

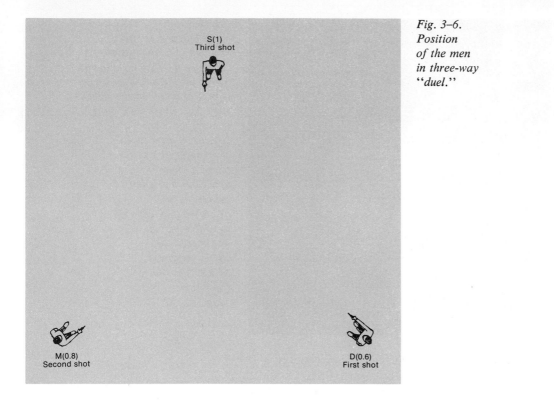

S(1)
Third shot

M(0.8)
Second shot

D(0.6)
First shot

Fig. 3–6.
Position
of the men
in three-way
"duel."

M will certainly shoot at D. Thus, the best strategy for D is to aim well away from both M and S.

Now it is M's turn. He certainly won't aim at D; he might succeed and S would then shoot him. And if M shoots at S, he seems to gain little: if he shoots S, D will certainly shoot at him; if he misses, S is likely to shoot at him with catastrophic results. Hence, one can argue M should also shoot well away from S and D.

Now S prepares to shoot. So far he knows D and M have both shot, but obviously not at him (he's heard no bullet whizzing by). He probably is in no danger from another round, since presumably D and M will make the same decisions again. On the other hand, if he shoots M, D will certainly shoot at him. On the basis of this reasoning, S fires far from M and D, and we start the cycle all over again.

If the preceding argument is correct, the shooting continues forever, with no one ever being hit. Of course, S must anticipate when M or D will tire of the suspense and decide to start aiming more purposefully; just before this time, S should shoot at M and take his chances with D. The selection of a desirable strategy (the optimization problem) depends not only on a study of the probabilities, but also on a guess as to the likely behavior of the other people in the contest.

This particular example has been used as a very elementary model for business competition among three companies. Then instead of bullets we can discuss price and product competition. In such a situation, it may be advantageous to everyone to have all three companies survive.

In this section, we have introduced some basic ideas associated with probability. We now wish to return to queues and see how these ideas are useful in doing experiments to measure the properties of queues.*

4 | QUEUEING STUDIES

In the study of queues, we might ask a variety of questions:

1. With a given level of business (average number of customers) and specified service facilities (e.g., number of toll booths), what is the average size of the queue?
2. How often will the queue be longer than a given length? How often does the barber lose possible customers who look in the window and then leave if there are more than two men waiting?
3. What is gained by adding another service station such as an additional toll booth or an extra barber and chair?
4. What can be gained by reducing the time required to service customers?

We would like to be able to answer such questions mathematically (or by computer studies), because in many cases we just cannot experiment on the actual system. We can hardly expect political approval for constructing another runway at an airport if we merely express the desire to experiment. We must be able to show that another runway will be helpful. Thus, queueing theory is concerned with seeking solutions to questions on the basis of studies of the model.

In this section, we will introduce the subject. In particular, we look at simple queueing problems. In more complex problems, we have to use the digital computer to simulate the system. That is, the computer operates as a model of the actual system. The following paragraphs consider primarily the question: What is the queue length in a very simple class of problems?

Arrival of Customers

In our attempts to simplify the problem, we first consider a very special way in which customers arrive at the service facility. We assume that they arrive completely at random and independent of one another. In other words, if on the average a customer arrives every 7 minutes, in any short interval T, the probability of a customer arriving is $\frac{T}{7}$. This is assumed true even if we look

*
Answer to problem in Section 3: 519 (the sum of 100, $\frac{100}{2}, \frac{100}{3}, \frac{100}{4}, \frac{100}{5}, \cdots \frac{100}{99}$, and $\frac{100}{100}$).

Fig. 3–7.
Lengthy queues build
up during a peak traffic
period at the tollbooths
serving the Triborough Bridge
in New York City.
(Triborough Bridge
and Tunnel
Authority)

at an interval just after the arrival of a customer. All we are saying here is that the customers arrive at a certain average rate. When they arrive is pure chance. Occasionally several customers arrive in close succession; on other occasions, there are long intervals between arrivals.

Now we can do an experiment to simulate a possible arrival of customers. We are interested in representing the arrival of customers at a store after it opens at 9 A.M. For the next hour, the customers arrive at an average rate of one every 4 minutes. We use the following steps:

1. We first divide the average 4 minutes into a number of smaller intervals. We might select one-minute intervals. The shorter the interval, the more accuracy we obtain, but also the more work we have to do. We should use an interval $\frac{1}{4}$ or less of our average time.

Fig. 3–8.
Possible arrivals
(average spacing
is four minutes).

2. We now look at each one-minute interval in turn. The probability of an arrival in this interval is $\frac{1}{4}$ because the interval is $\frac{1}{4}$ of the average arrival time. We first want to determine whether there is an arrival in the first interval, then the second interval (9:01 to 9:02), and so on for the hour.

3. We look now for some random experiment where the favorable outcomes are $\frac{1}{4}$ of all equally probable outcomes. If we have a deck of cards, we shuffle thoroughly and pick one card. We decide beforehand that if it is a spade (probability $\frac{1}{4}$), there is an arrival. If it is any other card, no arrival.

   Alternatively, we can roll two dice. The probability of rolling a 3 or a 4 or a 5 is $\frac{1}{4}$.* Hence, we roll the dice. If a 3, 4, or 5 comes up, there is an arrival; if 2, 6, 7, . . . , or 12 comes up, no arrival.

4. We repeat the experiment for each of the 60 intervals during the hour in which we are interested.

When we perform this experiment, we might find an arrival pattern as in Fig. 3–8.† Each vertical line represents the arrival of one customer. It is apparent that there are long periods when nobody arrives, for example, from 9:01 to 9:13 and from 9:47 to 10 A.M. Likewise, there are periods when customers are arriving one right after the other, particularly from 9:35 to 9:41.

Actually our way of determining the arrivals is not quite correct, since we do not allow more than one arrival in any one minute (and we could never have more than 60 arrivals). The graph does, however, indicate the idea of *random* arrivals.

This idea of random arrivals is often used in queueing problems because the problem is thereby greatly simplified. Actually, however, there are many cases in which the assumption is very close to the actual situation. Telephone calls during a busy period may be placed at random; customers do often arrive at a barbershop nearly at random. When highway traffic density is not too great, cars may arrive at a toll station according to our assumption.

Servicing Time

We now have the picture of customers arriving at random at our service facility. Each of these customers must be waited on,

*
When two dice are rolled, the probabilities of the various sums are:

Sum	Probability	Sum	Probability
2	$\frac{1}{36}$	8	$\frac{5}{36}$
3	$\frac{2}{36}$	9	$\frac{4}{36}$
4	$\frac{3}{36}$	10	$\frac{3}{36}$
5	$\frac{4}{36}$	11	$\frac{2}{36}$
6	$\frac{5}{36}$	12	$\frac{1}{36}$
7	$\frac{6}{36}$		

These numbers can be found easily. If one die were white and the other red, (so we can recognize which is which), there are 36 equally likely outcomes:

White 1	and	Red 1
White 2	and	Red 1
	Etc.	

Of these results, three give a sum of four (White 1, Red 3; White 2, Red 2; and White 3, Red 1). Therefore, three of the 36 equally likely outcomes are favorable if we want a four. The probability of a four is $\frac{3}{36}$.

The probability of a 3 *or* 4 *or* 5 is just the sum of the three separate probabilities, or $\frac{9}{36}$.

†
The data in Fig. 3–8 were determined by the author by rolling two dice. Each time a 3 or 4 or 5 appeared, it was interpreted as the arrival of a customer. Every other number rolled corresponded to the passage of a one-minute interval with no customer arriving. Merely by accident there were 15 arrivals, exactly the average per hour.

Dice are easier to use than cards. The trouble with cards is that the complete deck should be thoroughly shuffled before each draw of a card. Not only is this tiresome, but it is impossible for a human being to shuffle a deck thoroughly. There always is a tendency for groups of cards to stay together.

or *serviced*. Next, we assume that there is only one service facility: one runway, one barber, or one toll collector. Furthermore, we assume that the times required to service the various customers are either:

  a) all equal, or
  b) vary in exactly the same way as the inter-arrival times of the customers.

Assumption (a) means every customer takes the same time. This is approximately true for airplanes landing or haircuts.

In (b), the service times change from customer to customer. One man is serviced rapidly, another takes a long time. In a queue waiting at a bank teller's window, one customer may just be cashing a check. Another customer is bringing the day's receipts from a store and a bagful of coins must be counted.

Actually, we could complete this section by considering only the case of equal service times. We add the random variation to show the effect of different customer service times, an effect with which we are all familiar as a result of waiting in line behind the sweet little old lady who has a thousand questions she must ask.

Utilization

Before considering the problem of the queue length, we need to ask one question: Can the service station do the job of handling the arriving customers? In other words, is the service facility, with its specified servicing time, able to process the customers with their average arrival rate? Or, does the system break down completely with the number of waiting customers tending to grow and grow?

The answer to this question depends only on a ratio which we call β:

$$\beta = \frac{\text{average servicing time}}{\text{average time between arrivals}}$$

If β is less than one, the service facility can process the customers. A β greater than one means the system cannot work and the concept of a queue is meaningless. In other words, on the average the service facility must be able to process customers faster than they arrive on the average.

This result is hardly startling. If airplanes arrive at an airport at an average rate of two per minute and they can land on the single available runway at a maximum rate of one per minute (corresponding to a servicing time of one minute per customer), it is clear that the system cannot operate successfully. The sky will gradually fill with aircraft.

The quantity β, which is called the *utilization factor* in books on queueing, is a measure of the fraction of time the service facility is used. Thus, if $\beta = 0.7$, the service personnel and equipment are being utilized 70% of the time. In an eight-hour day, the barber, for example, would be giving haircuts 5.6 of the hours, on the average.

Length of the Queue

We are now ready to discuss the queueing problem, and in particular the average length of the queue. We have considered the arrival of the customers, the servicing time, and the utilization factor β. What would be nice now would be a simple mathematical formula for the length of the queue. Indeed, there are such formulas. We are assuming the arrivals are random and the servicing times either fixed or random. There is a formula for the average queue length in each case:

Fixed service times: $\dfrac{\beta}{1-\beta}\,(1-\tfrac{1}{2}\beta)$

Random service times: $\dfrac{\beta}{1-\beta}$

Unfortunately, these formulas aren't of much help except as very general guides. They are formulas which describe the average queue length after very long periods of time have elapsed. For example, we may require several thousand arrivals before we begin to approach the average given by the formula. In a barbershop, supermarket, or airport, we essentially start the problem over again every morning and arrival rates change during the day. Formulas which work only after hundreds or thousands of arrivals are not much help.

If mathematics does not help us, we use experiments to find the typical way the queue changes with time. Consequently, we return to our store example, where we had found arrivals at

① 9:01	④ 9:19	⑦ 9:35	⑩ 9:38	⑬ 9:41
② 9:13	⑤ 9:27	⑧ 9:36	⑪ 9:39	⑭ 9:44
③ 9:15	⑥ 9:31	⑨ 9:37	⑫ 9:40	⑮ 9:47

Let us assume that each customer requires 3 minutes for servicing. We can plot the queue length during the hour as shown in Fig. 3–9.

This graph is derived as follows. At 9:01 a customer arrives and is serviced until 9:04. At 9:13 another arrives who is serviced until

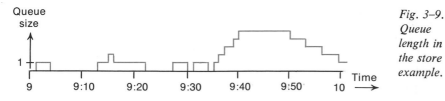

Fig. 3–9. Queue length in the store example.

9:16. At 9:15 another customer arrives and has to wait until 9:16 before his 3-minute service period starts. At 9:19 he is finished, but customer 4 arrives and is serviced until 9:22.

At 9:35 the seventh customer arrives. Thereafter, it is easier to calculate who is being serviced.

9:35— 9:38	Customer 7
9:38— 9:41	Customer 8
9:41— 9:44	Customer 9
9:44— 9:47	Customer 10
9:47— 9:50	Customer 11
9:50— 9:53	Customer 12
9:53— 9:56	Customer 13
9:56— 9:59	Customer 14
9:59—10:00	Customer 15

Customers who arrive before they are serviced must wait in the queue (the store services on a first-come, first-served basis). Consequently we can finish the graph by determining just after each minute who is still waiting or being serviced. If we do this, we obtain Fig. 3–9.

Several features are interesting. First, there is nobody waiting or being serviced 17 minutes of the hour. In other words, the service facility is being used only $\frac{43}{60}$ or 72% of the time. This number would, over a long time interval, be very close to 75% or β. Thus the utilization factor β indicates what fraction of the time the service facility is busy.

Second, even though the service facility is busy less than 75% of the time, there is another long period during which *five* customers are in the queue—one being served and the other four waiting. We might even calculate how long each customer has to wait by simply comparing his arrival time and the time when he starts being serviced.

Customer
① −0	④ −0	⑦ −0	⑩ −6	⑬ −12
② −0	⑤ −0	⑧ −2	⑪ −8	⑭ −12
③ −1	⑥ −0	⑨ −4	⑫ −10	⑮ −12

These times or the long queues may be intolerable. We may lose customers to a competitor.

Finally, we can calculate the average queue length during this hour. The graph on page 107 shows that, if we count a customer being serviced as a queue of one, the 60 minutes is divided as follows:

Queue of zero	17 minutes
Queue of one	19
Queue of two	5
Queue of three	5
Queue of four	4
Queue of five	10

The average queue length during this period is then

$$q = \frac{17 \times 0 + 19 \times 1 + 5 \times 2 + 5 \times 3 + 4 \times 4 + 10 \times 5}{60}$$

or

$$q = 1.8$$

On the average, we find 1.8 customers in the store during this period.

Comments

In the preceding example we studied the queues in a service facility which opens at 9 A.M. Customers arrive on the average every 4 minutes. Every customer is serviced in exactly 3 minutes. You remember that by an experiment such as throwing dice or drawing cards from a shuffled deck we could simulate the system. From this experiment, we determined the queue length during the hour from 9 A.M. to 10 A.M. Several comments should be made.

1. To study the store operation, we should repeat the experiment a number of times. Any one "day" may not be typical at all.

2. We can easily make the service times random instead of constant. Then for each customer we find experimentally the time he requires for service. The procedure is exactly the same as the way we found the times between arrivals. We find that this added randomness increases sharply the queue length.

3. We can study queueing problems with more than one service channel. We need only decide how the customer selects the channel he enters. He might, for example, always go to the channel with the shortest queue.

4. We can study experimentally the effect of reducing service time. In our store experiment, for instance, a cut in the service time to 2 minutes would cut the average time in the queue from 6.6 to 4 minutes and the average queue length from 1.8 to 1. These are the numbers which result when we use the same times as above.

5. We can study the advantages of an express check-out counter or an exact-change gate on a parkway toll plaza. In this case, we find the service times for all customers. We take those with short service times and process them through a separate channel.

Queueing problems are interesting for several reasons. They confront us continually in life. They show so dramatically the way randomness can affect system behavior. Finally, they can be studied experimentally and, indeed, experiments are often more revealing than advanced mathematical analysis. In complex studies, we of course have to use computers to perform the calculations. For ex-

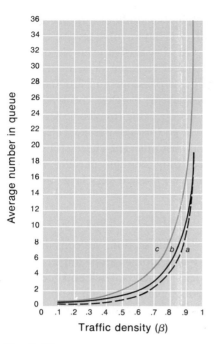

Fig. 3–10.
If the queues are too long, there are several ways to improve the system. The three curves show the mathematical formulas for queue length versus β in different cases: (a) Service time is halved, (b) Two servers work from one queue, and (c) Two queues are used, with customers sent randomly to one or the other.

The curves show it is best to reduce service time. Using two servers and one queue is almost as good. (Adapted from "Queues," Martin A. Leibowitz. Copyright © 1968 by Scientfic American, Inc. All rights reserved.)

ample, the study of airplane landing and take-off queues at a busy airport may involve several customers a minute.

An impressive example of queueing is the design of an international terminal at a modern airport. The facilities are built to process passengers in less than one hour, including (for departing passengers) baggage checking (perhaps 35 seconds/passenger) and ticket checking (maybe 65 seconds/passenger). Arriving passengers must go through passport clearance, health check, baggage pickup, and customs. The successful operation of the terminal depends on providing enough parallel channels and holding down service times to prevent very large queues.

5 | GAMES

In this section, we consider games as a new class of problems. We first look at an example.

Two friends, Dick and Joe, decide to play a game. At an agreed upon signal, they each push forward one or two fingers. To make the game interesting, they agree ahead of time that there will be a payoff after each play. The following rules determine the payoff:

If they match (both one or both two), Dick wins 10 cents.
If they don't match, Dick loses 10 cents.

These rules can be summarized in a table as shown below. Along the left, we list Dick's two possible strategies (One or Two).

| | | JOE | |
		One	Two
	One	10	−10
DICK			
	Two	−10	10

Along the top are listed Joe's strategies (One or Two). There are now four boxes. In these we place the *payoff to Dick* from each combination of strategies by Dick and Joe. Notice we arbitrarily decided to focus attention on Dick, and we list the payoff to him. What Dick wins, Joe loses, so the payoffs to Joe would be just the negatives of those shown.

The table above is called the *payoff matrix* of our game. It includes the two players, the strategies for each, and the payoff from each possible combination of strategies.

Dick and Joe play for awhile, with each selecting one or two fingers to show. Neither follows a regular pattern, so neither can predict his opponent's moves with any success. After playing all morning, they find their fortunes have varied. For awhile Dick was ahead, then Joe. At lunch, Joe owes Dick 20 cents.

Walking back from lunch, Dick comments, "The game seems a little boring. Let's try to introduce some excitement. Suppose we change the payoffs. If we match, I'll still win 10 cents. If I have

One and you Two, I'll pay you 20 cents. If I have Two and you One, no one will pay at all."

Joe agrees something must be done to make the game more interesting. As he understands, the payoff matrix now has the form:

		JOE	
		One	Two
DICK	One	10	−20
	Two	0	10

Joe thinks the change is trivial and actually wishes they would try a new game. But he agrees and they start playing again.

Dick asks if Joe would mind if he rolled a pair of dice before each play, just for diversion. Joe is now convinced Dick is foolish, but decides to humor him. They spend the afternoon playing.

By quitting time, they have played 800 times during the afternoon. Joe owes Dick $20.40. As Joe leaves for home, he wonders whether he has just had a streak of bad luck, or is there something about the game he doesn't understand?

We can "solve" this game. That is, we can find the best strategy for each player and also the average payoff expected each time the game is played. As you may already have noticed the game is really loaded in Dick's favor.

Are there other similar games that can be solved? For what class of problems can we find a solution?

The problems we are considering here make up the subject of *game theory*. A more accurate description is the theory of *two-person, zero-sum, finite* games. What does each of these italicized terms mean?

The games are *two-person*. That is, there are two players competing with each other. Furthermore, both players are intelligent. Both have the same picture of the payoff matrix. And each knows the other is intelligent.

The games are *zero-sum*. What one player wins, the other loses. There is no chance for the two players to cooperate so they both win. We have strictly a competitive situation.

The games are *finite*. Each of the two players has only a certain number of possible strategies. For example, we might have a 3 × 4 game: one player X has three strategies, the other player Y has four. In this case, the payoff matrix might look like

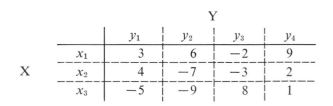

		y_1	y_2	y_3	y_4
	x_1	3	6	−2	9
X	x_2	4	−7	−3	2
	x_3	−5	−9	8	1

If player X chooses strategy x_2 and Y chooses y_3, the payoff to X is -3. X loses 3, Y wins 3. The entries in the payoff matrix are the payoffs to the player listed at the left.

Solution of Simple Games

When the game is described by only two possible strategies for each player, there is a simple algorithm for the solution.

We first look at the game to see if it can be simplified. Is there one strategy for either player which is obviously better than the other? For example, in the game

X obviously should always play x_1. No matter what Y does, X will win more than if he played x_2.

Now that we know X will always play x_1, let's see what Y should do. Y is intelligent and knows X is going to play x_1. Then Y certainly should play y_1. Thus, if both players are intelligent, each time the game is played X will win 4. We call the average amount X wins from each play the *value* of the game. Here the value is 4.

Similarly, in the game

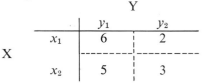

Y should always play y_2. Then X should play x_2. The value of the game is 3.

If neither X nor Y has a single best strategy, we can find the way each should mix his strategies. First consider X in the game

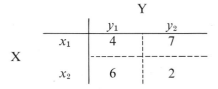

1. Subtract the second column from the first

$$4 - 7 = -3$$
$$6 - 2 = 4$$

We have the two numbers

$$-3$$
$$4$$

2. Reverse these numbers and throw away minus signs. We have

$$4$$
$$3$$

3. These represent the relative number of times X should play x_1 and x_2, respectively. In other words, $\frac{4}{7}$ of the time he should play x_1. In a single play, he selects x_1 or x_2 at random with probabilities $\frac{4}{7}$ and $\frac{3}{7}$.

We find Y's strategy by the same procedure. Subtracting the second row from the first gives

$$-2 \quad 5$$

Interchanging and discarding the minus sign gives

$$5 \quad 2$$

Y should play y_1 $\frac{5}{7}$ of the time, y_2 $\frac{2}{7}$ of the time.

The value of the game can be found by considering X's best strategy against either move by Y. If we pick y_1, X can expect to win 4 in $\frac{4}{7}$ of the time, 6 in $\frac{3}{7}$ of the time. The expected winning for X is then

$$4 \times \frac{4}{7} + 6 \times \frac{3}{7}$$

The value of the game is $\frac{34}{7}$.

The preceding description of the algorithm is not very satisfactory. We have not explained where the algorithm comes from. Indeed, the algorithm only works for two-strategy games. If each player has three or more possible strategies, we have to expand the algorithm with several additional algebraic steps. A more complete discussion would lead us into mathematics which is not of importance in the remainder of the text.

Thus, the algorithm for the 2×2 game is included here merely to describe the sort of solutions which we can obtain for games.

With the algorithm, we can solve our original game. Dick and Joe started by playing the game:

		JOE	
		One	Two
	One	10	-10
DICK			
	Two	-10	10

Application of the algorithm shows each player should select One or Two at random, each half the time. The value of the game is zero. If they play long enough, the average (per play) payoff should approach zero.

After lunch they switched to the game:

Now the best strategy for Dick is to play One $\frac{1}{4}$ of the time, Two $\frac{3}{4}$ of the time. On any single play, he throws the dice. If a 3 or 4 or 5 comes up, he plays One. If any other sum appears on the dice, he plays Two. (The dice are used merely to assure a random selection. He wants to avoid following a pattern which Joe might detect.) If Dick follows this strategy and Joe plays One, Dick will win ten $\frac{1}{4}$ of the time, zero $\frac{3}{4}$ of the time. The value of the game is 2.5 cents. If they play 800 times, Dick can expect to win about 800×2.5 cents, or \$20.

Comment

The examples of this section serve as only an introduction to games. Game theory started during World War II with studies of different tactical choices. During the last two decades, many economic and social situations have been studied. Current emphasis in research is on trying to find ways to solve games in which both players can profit if they adopt cooperative strategies. The hope is to apply game theory to social science problems in which we are looking for a strategy which, for example, will benefit both the power company seeking to locate new generators and the groups interested in improving the environment.

6 | LINEAR-PROGRAMMING PROBLEMS

An ice cream plant can make two flavors, vanilla and chocolate. The plant capacity is 1000 quarts per day, and the sales department says that it can sell any amount of vanilla up to 800 quarts and any amount of chocolate up to 600 quarts. If the profit per quart is 10¢ for vanilla and 13¢ for chocolate, what is the most profitable daily production?

Since the profit per quart is larger for the chocolate than for the vanilla, we should produce as much chocolate as we can sell. This would set the production of the plant at 600 quarts of chocolate. All excess production could then be applied to the vanilla ice cream, to add to the profit. The daily production of the plant is thus 600 quarts of chocolate and 400 quarts of vanilla, not a very difficult problem. We now solve this problem in a systematic manner because we want a technique which can work for more complicated problems.

The type of problems in the remainder of this chapter are called *linear-programming* problems. One of the earliest famous problems was a study in 1945 by George Stigler of Columbia University. He was looking for the cheapest diet which would provide the daily needs of calories, proteins, calcium, iron, vitamin A, thiamine, riboflavin, niacin, and ascorbic acid. He considered 70 possible foods.

The lowest-cost diet was a combination of wheat flour, cabbage, and hog liver. By mixing appropriate amounts of each of these, a man could live in good health for $59.88 a year. Costs today are higher, so it might require 30¢ a day.

This original solution was limited because computers were not available. Too many constraints could not be handled. If we were to solve the same problem today, we could add constraints which would ensure that on occasional days the diet was a little more appetizing.

Inequalities

In order to find a general method of solution, we need to use mathematical statements called *inequalities*. The inequalities are:

$x > 20$	meaning	x is greater than 20
$x \geq 20$	meaning	x is greater than or equal to 20
$x \leq 20$	meaning	x is less than or equal to 20

In other words, an inequality describes a range of permissible values. For example, in the first example, the inequality is just a simple way of saying x can have any value larger than 20.

We often combine two inequalities into a single expression. For example,

$$0 \leq x \leq 30$$

means that 0 is less than or equal to x, and x is less than or equal to 30. In other words, x can have any value from 0 to 30. As an example, if we have $30 and we want to buy a radio, the money spent on the radio R can be in the range

$$0 \leq R \leq 30$$

If we are also spending T dollars to buy tickets for a movie,

$$0 \leq R + T \leq 30$$

We can now describe the ice cream problem in terms of inequalities. We let V be the number of quarts of vanilla to be produced in one day and C be the number of quarts of chocolate to be produced in one day. One thousand is the total number of quarts produced by the plant for one day.

The number of quarts of vanilla to be produced may not exceed 800, but may be less than that quantity. We express this by the double inequality:

$$800 \geq V \geq 0$$

The vanilla ice cream produced may be any amount between 0 and 800 quarts. Similarly the quantity of chocolate ice cream to be produced can be represented by:

$$600 \geq C \geq 0$$

The number of unknowns can be reduced from two to one by remembering that the total production must be 1000 quarts of ice cream. With this limitation, if V quarts of vanilla are produced, there can only be $(1000 - V)$ quarts of chocolate ice cream.

Our statements are thus:

$$0 \leq V \leq 800$$
$$0 \leq (1000 - V) \leq 600$$

At this point, we have a model for the decision problem. The model states that there are several constraints on the quantity of vanilla, V, produced each day:

$V \geq 0$ (The amount of vanilla must be positive or zero, we cannot have a negative amount of vanilla.)

$V \leq 800$ (We cannot sell more than 800 quarts of vanilla.)

$1000 - V \geq 0$ (The amount of chocolate must be positive or zero.)

$1000 - V \leq 600$ (We cannot sell more than 600 quarts of chocolate.)

The last two inequalities are hard to interpret. They can be simplified if we manipulate them so that V appears alone on one side (as in the first two). To do this, we need some rules on how to change inequalities.

We can add any quantity (positive or negative) to both sides of an inequality. For example, the inequality is

$$x + 10 \geq 2$$

We can subtract ten from both sides to obtain

$$x \geq -8$$

We can multiply both sides of an inequality by any positive number. If

$$3x \geq 24$$

then multiplying by $\frac{1}{3}$ gives

$$x \geq 8$$

If we multiply both sides by a negative number, we must change the direction of the inequality. That is, "less than" becomes "greater than."

With these rules, we can simplify the four inequalities for the ice cream problem. We started with:

$$\begin{cases} V \geq 0 \\ V \leq 800 \\ 1000 - V \geq 0 \\ 1000 - V \leq 600 \end{cases}$$

In the third, we add V to both sides. In the fourth, we subtract 600 and add V. Then we have:

$$\left.\begin{matrix} V \geq 0 \\ V \leq 800 \\ V \leq 1000 \\ V \geq 400 \end{matrix}\right\} \text{ Model}$$

The second of these is clearly stronger than the third, the fourth stronger than the first. That is, if the second is satisfied, the third is automatically o.k. Hence,

$$\boxed{\begin{matrix} V \geq 400 \\ V \leq 800 \end{matrix}} \quad \text{Final model}$$

Now that the model is determined, we can turn our attention to the *criterion:* what is to be optimized? In this problem, we wish to maximize the profit. The problem specifies that the profit is 10¢ for each quart of vanilla and 13¢ per quart of chocolate. Thus, if P is used to represent profit (in cents), we can write

$$P = 10V + 13C$$

Once again, C is equal to $1000 - V$. The criterion (profit) equation becomes

$$P = 10V + 13(1000 - V)$$

$$\boxed{P = 13000 - 3V} \quad \textit{Criterion} \text{ (to be maximized)}$$

Finally, we must consider the *constraints* before we start the optimization. In this particular problem, there is no constraint specified. A constraint might be that the owner never permits production of less than 500 quarts of vanilla because of his personal desire to cater to his vanilla customers.*

Solution

Now that the problem statement is complete we turn to the optimization of the solution. We must select V to satisfy the two inequalities of the model, and at the same time to maximize P. If we look at P, we see that the profit is reduced as V is increased. Therefore, we should pick V as small as possible.

*
One might argue, for example, that the factory limit of 1000 quarts/day of ice cream is a constraint rather than a part of the model. It makes no difference whether a given part of the problem is considered part of the model or a constraint. In this example, we put everything into the model.

The model shows that V must be 400. The solution is then to produce 400 quarts of vanilla and 600 quarts of chocolate. The resulting profit, found from the relation for P, is 11,800 cents.

7 | GRAPHING INEQUALITIES

In retrospect, we can only marvel at the last section. There we took a ridiculously simple problem which can be solved by inspection. We managed to make a complicated problem which involved the definition of a model and criterion and the manipulation of a set of inequalities. Of course we did this because we need the more complicated approach when we come to more difficult problems. In the next section, we will look at a problem which cannot be solved by inspection. Then we will need to find a model in terms of a set of inequalities.

Before we consider this more complicated problem, we have to look at graphs for inequalities. We will need to portray graphically what we mean by an inequality with two unknowns. For example, we might have

$$2x + y \geq 6$$

How do we show on a graph what this inequality means?

The two unknown quantities are x and y. Therefore, we try to show this graph in the $x - y$ plane (Fig. 3–11). It makes no difference which we choose to plot horizontally, which vertically. Here we show y vertically.

The inequality divides the plane in two parts. In one part, the inequality is true. In the other part, it is not. The dividing line is found by plotting the *equation* we have when the inequality sign is replaced by an equal sign.

In our example, the equation is

$$2x + y = 6$$

This equation is plotted as a straight line. On one side of this line, the inequality is satisfied; on the other side, it is not.

The easiest way to plot this equation is to find the two intercepts. When y is zero, the equation states x is 3. When x is zero, $y = 6$. Figure 3–12 shows the straight line through these two points. (Any two points can be used, or we can find one point and the slope.)

Once the line is drawn, the inequality is satisfied on one side of the line. Which side? We can take any point on either side and try these values of x and y in the inequality. For example, the origin (x and y both zero) does *not* satisfy the inequality. Then, we can add the words on Fig. 3–13. If we want to be sure of the above calculation, we can pick a second point on the other side of the line and see whether the inequality is satisfied or not. In our

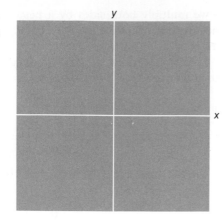

Fig. 3–11. The plane where we will show $2x + y \geq 6$.

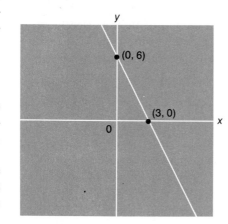

Fig. 3–12. Plot of $2x + y = 6$.

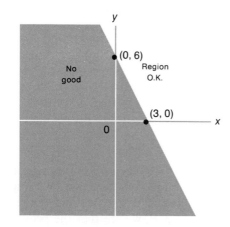

Fig. 3–13. Region for $2x + y \geq 6$.

example, we might pick an x of 3, a y of 6. We are above the line and the inequality *is* satisfied.

Actually, we usually use different terms than those of Fig. 3–13. Figure 3–14 shows the form. The region in which the inequality is satisfied is called the *feasible region*. We use shading on the *other* side of the line to show that this area is forbidden. The inequality is *not* good in the shaded region.

When we have several inequalities, each defines a line in the x-y plane, to one side of which the inequality is met. The total *feasible region* is then that area where *all* the inequalities are met.

An example is provided by the three inequalities:

$$2x + 4y \geq 12$$
$$x \leq 5$$
$$y \leq 3$$

The first inequality gives the allowable region shown in Fig. 3–15. The straight line passes through (6,0) and (0,3). The point (0,0) is not in the feasible region.

The second inequality allows anything to the left of $x = 5$. The third inequality admits all below $y = 3$. When these two lines are added, we obtain Fig. 3–16. The feasible region is the triangle.

With this background, we can now turn to a more interesting linear-programming problem.

8 | A TRANSPORTATION-PLANNING PROBLEM

The ice cream problem involves a search for a "best" solution to a problem for which more than one solution is possible. The "best" solution is one which achieves either a maximum or a minimum under the constraints that are imposed. These constraints are expressed mathematically as either *linear equations* or *linear inequalities*. Such problems have been named linear-programming problems.

By a linear equation here, we mean each variable, x or y, appears only to the first power. For example, $3x + 4y = 2$ is a linear equation. The term linear also refers to the fact that the plot is a straight line. In a similar fashion, a linear inequality is an inequality which becomes a linear equation when the inequality sign is replaced by an equal sign. Our examples of linear-programming problems in this chapter are purposely simple, that is, the number of variables has been limited. Practical problems frequently involve a large number of variables, often hundreds. Then high-speed computers must be used to find the solution.

It is fascinating to discover the variety of design and planning problems that involve linear inequalities, and once we learn to use linear inequalities, we can apply them to many different types of problems. In this section we treat a *transportation-planning problem*.

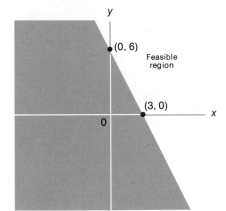

Fig. 3–14. Feasible region for $2x + y \geq 6$.

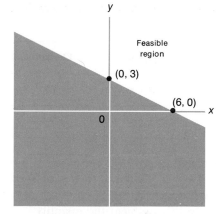

Fig. 3–15. Feasible region for first of three inequalities.

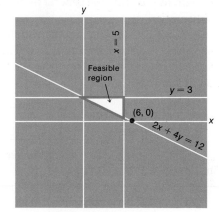

Fig. 3–16. Feasible region defined by all three inequalities.

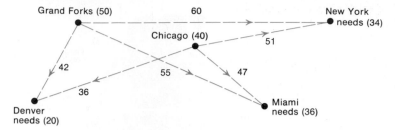

Fig. 3–17.
*Sources and needs of wheat
and shipping costs (numbers in
carloads and in dollars/carload).*

Statement of the Problem*

A grain dealer owns 50 carloads of wheat in Grand Forks, North Dakota, and 40 carloads in Chicago. He has sold 20 carloads to a customer in Denver, 36 carloads to a customer in Miami, and the remaining 34 carloads to a customer in New York. He wishes to find the minimum-cost shipping schedule, on the basis of the freight rates shown in Fig. 3–17 and Table 3–1. For

Example suggested by "Operations Research for Students of Business," notes by James R. Jackson, published by University of California at Los Angeles.

To From	Denver	Miami	New York
Grand Forks	42	55	60
Chicago	36	47	51

*Table 3–1.
Wheat freight
rates in dollars per
carload.*

example, it cost $42 per carload for shipments from Grand Forks to Denver. Different modes of shipment involve rates which are *not* proportional to the distance between the cities.

In this problem the quantity of wheat which has been sold is just equal to the quantity of wheat in storage. This need not necessarily be the case in practice, but we have designed this problem to illustrate the general method with the use of the simplest mathematics.

We can present the shipment data in a table, which indicates quantities as shown in Table 3–2. The figure in each box will be the quantity to be sent from the source at the left to the destination at the top.

↓ Quantity in storage	Quantity needed → To From	20 Denver	36 Miami	34 New York
50	Grand Forks			
40	Chicago			

*Table 3–2.
Data
arrangement for solving
grain problem.*

Our problem is to find the quantity of wheat to fit into each of the six empty boxes so that three conditions are met. These are:

1. The amounts in the first row add up to 50 and the amounts in the second row add up to 40 (the total amounts to be shipped from Grand Forks and Chicago, respectively).
2. The amounts in the first, second, and third columns add up to 20, 36, and 34, respectively (the amounts to be delivered to Denver, Miami, and New York, respectively).
3. The total freight cost is a minimum. This cost for each shipment is obtained by multiplying the amount of wheat in each box by the rate for that trip. The six products are then added to give total cost.

It is not enough to pick the numbers in each box so that the sums of the rows and columns are correct. This would be a feasible solution, but we must also minimize the total cost. With "cut and try" methods, we might find such a solution. However, a systematic approach usually takes less time. For problems with more shipping points and with more destinations, a systematic approach (an algorithm) and a computer are essential.

If we designate the quantity shipped from Grand Forks to Denver as x units, then $20 - x$ units must be shipped from Chicago to Denver to complete the order. Similarly, if we designate the amount shipped from Grand Forks to Miami as y, the amount from Chicago to Miami must be $36 - y$. Now the amount from Grand Forks to New York must be $50 - x - y$ if the total out of Grand Forks is to be 50. The amount from Chicago to New York must then be $34 - (50 - x - y) = x + y - 16$. We automatically satisfy the requirement that the total shipped from Chicago is 40, since the total amount sold equals the amount ordered.

Table 3–3 now represents the quantities of wheat to be shipped from each storage house (Grand Forks and Chicago) to each destination (Denver, Miami, and New York). Because of the way the six entries are selected, each row and column adds up properly. Our task in optimization is now to select x and y and, hence, each of these six entries:

$$x, y, 50 - x - y, 20 - x, 36 - y, x + y - 16$$

↓ Quantity in storage	Quantity needed → To From	20 Denver	36 Miami	34 New York
50	Grand Forks	x	y	$50 - x - y$
40	Chicago	$20 - x$	$36 - y$	$x + y - 16$

Table 3–3.
Table 3–2 with unknown quantities x and y defined.

There is one additional part of the model (or the constraints). Each of the entries in Table 3–3 must be *positive* or *zero*. We cannot ship negative quantities of wheat. Therefore, we must add the six inequalities:

$$x \geq 0$$
$$y \geq 0$$
$$50 - x - y \geq 0$$
$$20 - x \geq 0$$
$$36 - y \geq 0$$
$$x + y - 16 \geq 0$$

These inequalities can be rewritten in the following form, if we rearrange terms:

Derived by adding terms to both sides of the inequalities $\left\{ \begin{array}{l} x \geq 0 \\ y \geq 0 \\ y \leq -x + 50 \\ x \leq 20 \\ y \leq 36 \\ y \geq -x + 16 \end{array} \right.$

Determination of the Feasible Region

We now have the model represented by Table 3–3 and the six inequalities. Before considering the criterion function (the cost to be minimized), we can represent our six inequalities on a graph of y versus x. Each inequality states that we must work on one side of a straight line in this plane.

For example, the first inequality

$$x \geq 0$$

requires that we operate to the right of the y axis. The fourth

$$x \leq 20$$

places us to the left of the vertical line $x = 20$. The effect of these two inequalities is to restrict our choice of x and y to the region shown in Fig. 3–18.

Figure 3–19 shows the six straight lines and the corresponding permissible regions. The six inequalities require that we operate (select x and y) either within the feasible region indicated in Fig. 3–19 or *on the boundary of this region*.

Criterion Function

Figure 3–19 and Table 3–3, together, now represent the model and the constraints of the total problem. Before we can start the optimization, we need to determine an equation for the criterion, the cost. Optimization then consists of looking for a point (x and y) within the feasible region at which the cost is minimized.

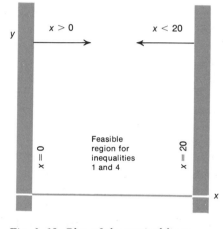

Fig. 3–18. Plot of the vertical lines x = 0 *and* x = 20.

Fig. 3–19.
Graphical representation of the six inequalities of the model.

The cost of each shipment in each of the boxes of Table 3–3 can be expressed by multiplying the number of carloads shipped times the cost of shipping for each carload. The sum of all these individual shipping charges then gives an expression which represents the total cost of the entire shipment. Thus, the shipment from Grand Forks to Denver is x carloads. Each costs $42. Then the cost for this shipment is $42x$. In a similar fashion, we can compute the cost of shipment between any two cities for each of the squares. The total cost of shipping is:

$$C = 42x + 55y + 60(50 - x - y) + 36(20 - x) + 47(36 - y) \\ + 51(x + y - 16)$$

If this expression is simplified,

$$C = 4596 - 3x - y$$

This C is the total cost of all shipments. It gives us the cost for any values of x and y we may choose.

Thus, the problem we started with now can be stated as follows. How do we pick values for x and y which

1. are in the feasible region of Fig. 3–19 or on the boundary, and
2. minimize C as given above?

Solution

After all this preparation, we are finally ready to solve the problem. There are two ways we can find a solution.

Trial and error. We can guess at x and y in the feasible region. We can find C. We then try different x and y values and again find C. We continue this until we have the least possible value of C.

This trial and error would be hopeless if we had to try every pair of x and y values in the feasible region. Fortunately, the solution always comes out at a vertex or corner of the feasible region. Figure 3–20 shows there are only six vertices. We only need to try these six pairs of x and y values.

Vertex	x	y	$C = 4596 - 3x - y$
A	0	36	4560
B	14	36	4518
C	20	30	4506
D	20	0	4536
E	16	0	4548
F	0	16	4580

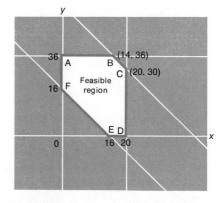

The minimum cost occurs at C. The problem solution is

$$x = 20 \qquad y = 30 \qquad C = 4506$$

Finding the x and y values at each vertex and calculating C is not difficult. Even in quite complex problems, a computer can

Fig. 3–20.
Feasible region with vertices shown.

be used to search for the best vertex. In very complex problems with hundreds of constraints, even computers take too long. Then there is a search procedure called the simplex method which allows you to program the computer so that we first guess one vertex, then always move to a vertex where C is less. We don't bother to look at vertices where C is larger.

Graphical. We can also solve our problem by a graphical approach. This procedure shows why the solution always is at a vertex. Let's return to our problem statement which has two parts:

The feasible region of Fig. 3–19
The C equation, $C = 4596 - 3x - y$

We can rearrange terms in the cost equation in the form

$$y = -3x + (4596 - C)$$

This is the equation of a straight line. The slope is -3. The intercept on the y axis is $(4596 - C)$. Each different value of C gives a different line. For example, if C is 4596, the line is $y = -3x$ (Fig. 3–21). We change C to 4580, the line moves as shown in Fig. 3–22.

If we look at Fig. 3–22, we see that as C decreases, the cost line moves parallel to itself upward to the right. The minimum C must occur when the line is in the position shown in Fig. 3–23. The best C occurs when the line passes through

$$x = 20 \qquad y = 30$$

There the equation for C tells us that

$$C = 4596 - 3(20) - (30) = 4506$$

The total solution for least cost is shown in Table 3–4.

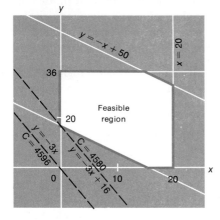

Fig. 3–21. Cost line ($C = 4596$) added.

Fig. 3–22. Cost line ($C = 4580$) added.

To From	Denver	Miami	New York
Grand Forks	20	30	0
Chicago	0	6	34

Table 3–4.
Minimum-cost
solution for wheat
problem.

It is interesting that *no* wheat should be shipped from Chicago to Denver even though this is the cheapest single route.

Figure 3–23 shows that the largest C occurs when the cost line is downward to the left as far as possible. Then C is 4580.

If this is a real problem, the differences between maximum and minimum cost do not seem very impressive.

Least cost: $4506   Greatest cost: $4580

All this work only saves $74. If the profit we can expect in running this business is only 3%, however, the profit expected from this sale is about $136. By using optimization, we have added $74 to this figure. This is hardly a trivial gain.

9 | FINAL COMMENT

In this chapter, we have continued the discussion of decision making. We have considered three kinds of decision problems: queues, games, and linear programming. In each case, our purpose is to introduce the reader to a class of decision problems. We try to see what are typical elements of the problems. There is no attempt to go deeply enough to permit the reader to solve any more than the simplest problems.

Thus, our major objective is to show that various decision problems can be approached intelligently and logically. Decision making need not be entirely a guess and a prayer. While no one should run a business strictly on the basis of these mathematical viewpoints, knowledge of problem-solving techniques can help the decision maker to gain a better understanding of his problem.

Let us consider queueing as an example. We hope that the reader will gain some appreciation of what a queue is, why queues form, and how queues can be avoided. Once we understand what a queue is, certain facts are important:

1. The system must be built so average service time is less than the average time between arrivals.
2. The randomness of arrivals and service times is really what causes queues.
3. Even though the average queue length may be small, there are frequent periods when the queue length is very large (possibly greater than 10).

Once we understand the general behavior of queues, we can see obvious actions that can be taken to reduce the queueing problem:

a) The arrivals can be made more regular. (Drive-in theaters have a playground to attract some customers early in an attempt to reduce queues just before show time.)
b) Service times can be made less random. Buses that require exact change need much less time to load passengers, and have the side advantage of discouraging robbery.
c) Service times can be reduced. Banks require the customer to fill out the deposit or withdrawal slips before he enters the line at the cashier's window.
d) Priority rules can be established. Airline passengers who already have tickets need not check in at the counter, but can go directly to the gate.

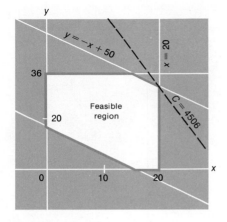

Fig. 3-23. Feasible region with minimum-cost line added.

Fig. 3–24.
Traffic congestion. (City of New York
Department of Traffic
photo by C. DeLuise)

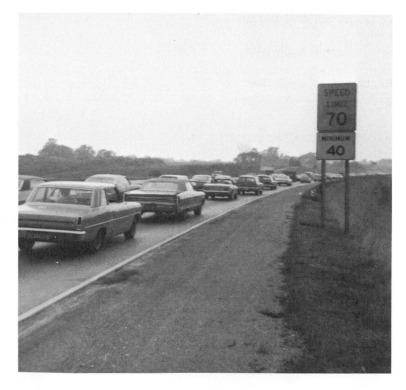

Fig. 3–25.
Congested highway
with minimum speed
optimistically
indicated.
(Photo by
Richard O'Leary)

Tunnel toll / Exit to Brooklyn / Interchange with Cross-Island Pkwy. / Interchange with Northern State Pkwy.

Manhattan Island — Bottleneck A — 3 to 8 Inbound lanes — Bottleneck B — Bottleneck C — Bottleneck D — Travel — Long Island

Fig. 3–26. A few bottlenecks on a main highway.

e) Additional service channels can be added. When bank queues become so long customers are irked, the alert manager quickly opens other cashier windows.

Finally, the chapter should point out the complexity of common problems. Often we find a system in which there are a sequence of service channels or bottlenecks. For example, the Long Island Expressway is the main highway from Long Island into Manhattan. The normal traffic congestion has given the 50-mile road the affectionate title of the "world's longest parking lot."

On a typical morning rush hour, there are several points at which severe bottlenecks occur (Fig. 3–26). Often enormous queues form at each of these points. Frequently, one queue backs up all the way to the preceding bottleneck, and traffic almost comes to a halt.

It is clear that reducing the queues at Bottleneck D may really not help the system at all. All this does is make the situation that much worse at C, B, and A. Indeed, a severe bottleneck at D (perhaps caused by a stalled car in the middle lane) may be just what the system needs. Drivers coming from east of D may have an unusually long wait as they approach D, but from there on in to Manhattan traffic flow may be very smooth.

In other words, bottlenecks may actually be desirable. Perhaps an inexpensive method of control for our highway would be to arrange for a stalled car or two each morning at appropriate locations.

In a variety of more significant problems, we sense that bottlenecks really may be advantageous. For example, the rapid development of computers during the past two decades led to the use of more than 75,000 computers by 1970. The resulting problems in business management, city operation, and many other fields can only suggest that we should not complain about the various bottlenecks which have slowed down this progress. This slowdown may have been essential because our social institutions change slowly.

Questions for Study and Discussion

1. "Most engineering design activity currently relies on engineering judgment." Mention some of the kinds of judgment on which people rely when they buy manufactured goods, for example, canned or frozen vegetables, electrical appliances, automobiles.

2. Independent testing laboratories have been formed to help consumers by checking on various products. If you were to start a small testing laboratory in your community how would you go about rating canned tomatoes? Hamburgers? Hot dogs?

3. More than half the people in the United States are under thirty, and the average age of all Americans is going down slightly each year. Suppose you inherit $20,000 and decide to start your own store. How would you do a market survey to decide what kind of store to open in your community? State your major criterion and the various constraints which you see.

4. What is meant by the term "feasible region" in the solution by graphical means of a linear-programming problem?

5. In developing the problem on shipping wheat (Section 8), the text shows the inequality $50 - x - y \geq 0$ rewritten in the form $y \leq -x + 50$. How would the graph of the feasible region look if the original inequality had been changed to $x \leq 50 - y$?

6. In taking a 25-question True-False test you pick T as the answer to the first three questions. If the correct responses are truly randomly distributed throughout the test what are the chances that the correct response to the fourth question is T?

7. What examples of queue formation in your school can you think of? (Consider, for example, classrooms just before or after class, science labs, cafeteria, gym, principal's office, library, bulletin boards.) In each case, is it a single-service queue or a multi-service queue?

8. a) If customers arrive at the teller's window in a bank at random, but at an average rate of one every 30 seconds, what is the average time between arrivals in terms of seconds? Of minutes? Of hours?

   b) If the teller always takes 20 seconds to service each customer, what is β?

   c) How long is the average queue length? (It is easier here to use common fractions rather than decimals.)

   d) Suppose the teller still averages 20 seconds per customer, but the lengths of individual transactions are random, does this change β?

Modeling
4

I | THE NATURE OF MODELS

According to an old story, six blind men who had never seen an elephant tried to decide what it was.* The first man, feeling the elephant's flat, vertical side, concluded that the beast was similar to a wall. The second man touched a round, smooth, sharp tusk and decided that an elephant is similar to a spear. Grasping the squirming trunk, the third blind man said that the animal resembled a snake. The fourth man, who touched a knee, observed that elephants resembled trees. From an exploration of the ear of the elephant, the fifth man was convinced that the animal had the shape of a fan, while an examination of the tail convinced the sixth blind man that an elephant was similar to a rope.

Each, of course, was partially correct, but insofar as a complete representation of the elephant was concerned, all were wrong. Each man, after observing the "real world," formulated a description, or *model*. But since the observations were incomplete, the models were incorrect.

Every time we describe an object, we are really making a model. We use our senses to find information about the real world. From this information, we decide what the object is. Then we pick out the *important* features. These make up the model.

The model is an efficient way of viewing things. A good model includes only those parts which are useful. But, by restricting our thoughts to a few features, we are able to understand the object or system. We can anticipate the effects of actions we might take. On this basis, we can select the best action. Thus, man's ability to control his environment and to build useful systems depends directly on his capacity to find models.

Models Are Usually Quantitative

Models begin as conceptions; they are ideas about the structure and nature of something. Before we can go very far with the model, we usually have to develop a *quantitative model*, one which tells how much, where, and when in terms of numbers. In other words, we use the language of mathematics to describe our situation.

* The Blind Men and the Elephant (J. G. Saxe, 1816–1887). The poem is given in Question 15 at the end of the chapter.

Fig. 4–1.
The girl is testing
her left eye. Every few
seconds, a point of light
appears on the screen. She
moves a lever in the direction
where she sees the light. From
this test, we can make a map of
the field of vision (the regions
where she sees the light).

This map is an example of a
model. Here we model the seeing
ability of the patient. From a
study of this model, the physician
measures the condition of the
eyes and can detect the
beginning of certain disease.
(IBM Corporation,
University of Texas
Southwestern
Medical
School at
Dallas)

The importance of a quantitative model can be illustrated by the task of the man who is in charge of scheduling elevators in a twenty-story office building from 8:45 to 9 A.M. every weekday morning as people arrive for work. There are six elevators, each able to stop at every floor.

When the building is first opened, the elevator supervisor simply loads each elevator in turn. As soon as it is loaded, it departs. Our supervisor notices that service is extremely slow. Occasionally, an elevator is loaded with one passenger for each of 15 different floors. The poor passenger who works on the highest floor has a ten-minute ride. (Especially since the passengers at the back of the elevator as it starts up always seem to be the ones who want to get off first.)

As the complaints about service become more and more bitter, our supervisor wonders if service might be improved by using a better plan. He might, for example, use two elevators for floors 2-8 only, two others for floors 9-15, and the other two for 16-20. He might stop only at even-numbered floors and force employees to walk down one flight of stairs to reach the odd-numbered floors. Obviously, there are many possible strategies which could be tried to improve service.

The choice of a desirable strategy depends on the way employees arrive in the morning for each of the floors. For example, the fifteenth floor holds the executive suite for the top officers of the company. If any of them should arrive during this busy period, they must be delivered as rapidly as possible. This priority for

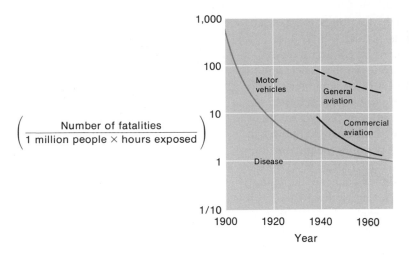

$$\left(\frac{\text{Number of fatalities}}{\text{1 million people} \times \text{hours exposed}} \right)$$

Fig. 4–2.
A different model which shows the willingness of people to accept risk. The vertical axis is a measure of the risk involved in different activities (the number of deaths in the U.S. divided by the total number of hours a million people are exposed to the risk). For example, 10 means that there are ten deaths every hour a million people engage in this activity.

The interesting part of this model is that just living (with possible disease) has about the same risk as driving a car or flying a commercial airplane. (Only about 5% of the people ever fly; 85% ride in a car.) People seem willing to accept this risk of 1. Cigarette smoking has a risk of 3, skiing about 1, and normal railroad travel 1/10. This model gives us some idea how safe we must make new forms of transportation (moving sidewalks or 300-mph trains).

The model is discouraging if we are interested in cutting down traffic accidents. It suggests that the average person may be willing to accept today's death rate. (Chauncey Starr, "Social benefit versus technological risk," UCLA report, 1969.)

Floor 15 means that at least one elevator should go directly to this floor. Similarly, certain floors may have many employees and should be given preference. In other words, to derive a sensible plan or strategy our supervisor needs at least a rough quantitative picture of the flow of people. This is the model.

The model may be very approximate and rough, guessing only that the number of employees is the same for each floor. Or it may be very detailed with the number for each floor, their probable arrival times, and their relative importance to the company. If the model is very simple, the supervisor may be able to decide intuitively on a strategy; if the model is very detailed, a computer may be required to evaluate and compare the possible strategies.

The important feature of this example is that no intelligent decision is possible unless a model is used, whether or not the supervisor calls his picture a model.

2 | THE GRAPH AS A DESCRIPTIVE MODEL

To form a model, we need to collect data about some aspect of the real world. We might wish to determine whether a simple relationship exists between the heights and the weights of twenty-year-old men. After making several experimental measurements, we may secure a set of related numbers such as 5′6″, 130 lb; 6′1″, 180 lb; 5′7″, 155 lb. But it is difficult to discover any systematic relationship in this way. Even though we may have a reasonable expectation that as height increases, weight will increase, this verbal model is vague and imprecise.

In order to present the data in a way which we can interpret more easily, we make a graphical plot, as in Fig. 4–3. Each point represents the height-weight data for one man. We notice now that the points are not scattered, but seem to be closely grouped. What can we say about the relationship between height and weight?

Fig. 4–3. Height-weight data for 20-year-old men.

A straight line can be drawn through the points to represent averages for these data (Fig. 4–4). This picture of the data is a *graphical model* of the relationship between height and weight.

This graphical model presents a clear but simplified description of the real world. It can be used as the basis for some reasonable predictions. From the straight-line average we can estimate the probable weight of a twenty-year-old man even though we know only his height. Let us suppose that we wish to estimate the weight of someone who is 73″ tall. The corresponding weight for this height can be immediately obtained from our graph. The graphical model permits us to estimate the weight of such an individual, even if our original data did not include any individual of this height. Thus, we may predict from our graphical model that an individual who is 73″ tall probably weighs about 180 pounds.

Fig. 4–4. Weight expected from a man 73″ tall.

An Equation As a Model

While the graph of Fig. 4–4 is an appropriate model for our weight-height relationship, it is also often convenient to have a mathematical equation. The straight-line graph is exactly equivalent to an equation, which in this case happens to be

$$W = 8H - 407$$

(We derive this in the next paragraph.) Even though the graph and the equation say exactly the same thing, it is often a nuisance to have to redraw the graph every time we want to tell someone what the model is. Furthermore, it is frequently easier to work with the equation. For example, what is the height expected for a 150-pound man? From the equation,

$$150 = 8H - 407 \text{ or } H = 70''$$

Since our graph is a straight line, we know from algebra that the equation has the general form $y = mx + b$, or in this case,

$$W = mH + b$$

where W is the weight, H is the height, m is the slope, b is the vertical axis intercept. To complete the equation, we must therefore determine the values of the constants m and b. Since b represents the W, or vertical intercept of the line, Fig. 4–5 reveals that the line cuts through the W-axis at a value of -407. Hence,

$$W = mH - 407$$

We can measure the slope m directly. When H increases by 10 inches (for example, from 60 to 70), W increases by 80 pounds. The slope is thus $\frac{80}{10}$ or 8, and the equation is

$$W = 8H - 407$$

Once the equation is determined, the weight W can be found for any given height, or the height for any given weight.

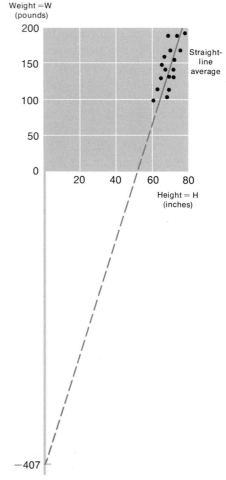

Fig. 4–5. Straight line extended until it hits the W axis.

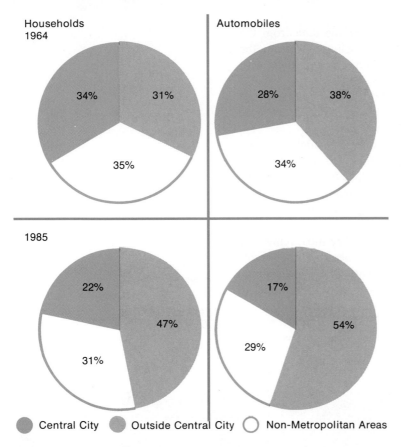

Households
1964

Automobiles

34%	31%
	35%

28%	38%
	34%

1985

22%	47%
31%	

17%	54%
29%	

● Central City ● Outside Central City ○ Non-Metropolitan Areas

Fig. 4–6.
A completely
different form
of graphical
model. In this case,
we are interested
in showing the changes
expected from 1964 to
1985. The "pie" charts show
how the suburbs will grow,
primarily as the central city
becomes less important (if
present trends continue).
Thus, a model can be
portrayed in many
different forms.
(Technology
Review *of*
M.I.T.)

We can use this model to predict points which were not originally in our data sample. But these predictions must be carefully examined. For instance, the line we have drawn tells us that an individual who is 54″ tall will weigh about 25 lb; worse yet, a 24″ person can be expected to tip the scale at an impressive *negative* 215 lb! Surely these are curious figures for twenty-year-old men. What is wrong with our model?

The straight-line average that we drew was extended so that the *W* intercept (-407) could be found. But we are not entitled to say that all points on the entire line must represent real situations. Actually, the only use we can make of this model is to predict within our data field. We know that anywhere inside of the cluster of measured points we are in the neighborhood of a real-world possibility. However, *we run into the danger of unrealistic prediction if we apply the model beyond the region that has been measured.* This danger applies to either the graph or the equation.

Our model, therefore, has its limitations. It must not be used to predict beyond the region of results experimentally obtained unless there are very good reasons to believe that real-world laws are not being violated. To test this model requires that we obtain a fresh sample of twenty-year-old men, either the data from the same school for other years or from other schools or from the army. Then we either enter them on the plot or compare them with predictions made from the algebraic model.

The equation is a mathematical model which says exactly the same things as the graphical model. Both of these models are more useful than the verbal model with which we started.

One of the most interesting aspects of this model is that it turns out to be so simple. This is quite unexpected. If we think of the people we meet while walking down a busy city street, we know that all sizes and weights are combined. It is true that the sample studied was passed through two strainers (age and sex) to make it more manageable.

We should give one final word of caution. We must remember where our original data came from. If we had measured the heights and weights of the members of the Kansas City Chiefs professional football team, we probably could not expect to use the model to predict the weight of a six-foot, 25-year-old starving artist. At least, we should be suspicious of the prediction. The model is only useful as long as we stay with the part of the world from which the model was derived.

Fig. 4–7.
City traffic congestion is not a new phenomenon. This picture was taken in 1905 at the corner of Dearborn and Randolph Streets in Chicago. (The Bettmann Archive)

3 | A DESCRIPTIVE MODEL OF TRAFFIC FLOW

A key urban problem is the question of how to handle motor traffic in the streets. Some cities have gone so far as to ban all automobiles from a few streets. This does not so much solve the problem as eliminate it—at least from those streets. Furthermore, it sometimes substitutes other problems, for example for the elderly and infirm, or for the shopkeeper trying to attract customers. It also increases the cost of goods since delivery expenses rise.

If a traffic engineer is to improve present conditions he must be able to predict the results of changes. To do this he must construct a model of the traffic flow in and around the city. Since such a model is too complicated, we use a study of the simpler circumstances within a school; even this we limit to what goes on at a single corridor intersection, such as that shown in Fig. 4–8. The limited model we derive for a single intersection could be extended to a whole building. The resulting larger model can be of practical use to school administrators, schedule-makers, and architects.

In order to construct our model, we must determine what affects the behavior of the system. We are primarily interested in the rate at which people (including teachers and custodians) pass from one corridor to another, in other words, in the density of traffic as measured in people per minute. The measurement will be made by *sensors*, devices which respond in some way whenever a person passes. An example would be the device often called an electric eye, but for short-term service a much more practical sensor is a person stationed at the proper spot.

Figure 4–9 shows the intersection with measurement points identified. Traffic problems usually arise at intersections rather than in the main corridors.* What we need for the present is a count, minute by minute during the time when classes are being changed, and for a minute (or more) before and after.

Table 4–1 shows a possible result of such a set of measurements. The actual number of people who go through the intersection during the counting period is about half that shown because each person was counted twice, once when he entered and once when he left (if we forget people like Joe and Susan).

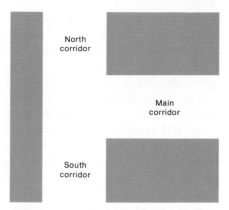

Fig. 4–8. A school corridor intersection.

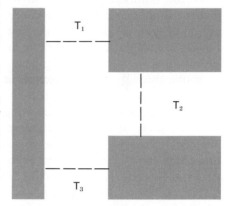

Fig. 4–9. Where traffic density is measured.

* As Joe passes position T_2 going east, he sees that Susan is going south past T_1. He reverses his field and meets her in the middle of the intersection. Therefore, they both hold up traffic and he is counted three times at T_2; he might even stroll a few yards down the south corridor with Sue and add two more tallies to his total.

Minute number	Counting station			Total counts
	T_1	T_2	T_3	
1	26 counts	50 counts	30 counts	106
2	42	63	47	152
3	61	102	55	218
4	112	184	73	369
5	38	42	28	108
6	22	17	9	48
			Total for period	1001

Table 4–1. Detailed traffic count at one intersection, end of first period.

The data can be shown more effectively if we present them in a plot, rather than a column of numbers. One form of plot which is often used in displaying a count of events is called a histogram. It has the form shown in Fig. 4–10 for the total number of individuals passing through this intersection.

The height of each vertical column is proportional to half the total count for the minute indicated below it, half the total for the reason just explained. It is evident that most of the traffic, but not all, occurred during the interval between classes. The school administrator confronted with such a bar graph might well be suspicious of the large traffic during the minute before the bell rang for class changing: Are some teachers dismissing their classes early? He might also be disturbed by evidence of a good deal of tardiness.

Further information, useful to the scheduling officer, can be obtained if another bar graph is made, this time of total counts at the end of each period during the day (Fig. 4–11). Here the peaks shown for traffic at the ends of the first and fourth periods might suggest altering the room assignments in such a way as to lessen the traffic through this intersection at these times. For example, more students might be scheduled for successive classrooms in the same corridor to avoid the intersection.

A full study of a school's traffic pattern requires that data be obtained at every intersection for every class-changing period throughout the day or even the entire week. If there are ten important intersections, the total data can be portrayed in fifty histo-

Fig. 4–10.
Bar graph of traffic
at one intersection at the
end of the first period.

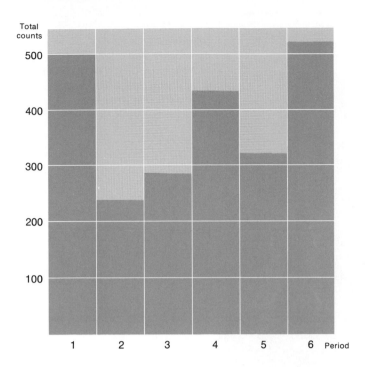

Fig. 4–11.
Bar graph
of total daily traffic
at one intersection.

grams similar to Fig. 4–11 (ten for each of five days) or similar curves on a minute-by-minute basis. Inspection of these data then reveals the times and locations of major congestion.

In processing data of this type, it is common practice to indicate those particular data points which are either unusually large or notably small. We might show in red all parts of the histograms which are larger than a predetermined amount (corresponding to troublesome congestion). We may choose to make up a small separate list showing critical data only. Such selective presentation of critical data is called "flagging"; we flag data which are especially important.

This type of traffic-flow study is closely related to the queueing problem described in Chapter 3. Queues occur when traffic density increases. In the city traffic problem, the results are obvious. Excessively long travel time ruins emergency services (ambulances, fire trucks, and police cars), lowers the income of taxi drivers, raises the cost of goods, and causes lost time for workers. One of the reasons some small businesses are moving out of the cities to the suburbs is the consistent tardiness of workers because of transportation problems. In one recent study, a manufacturing business employing 700 workers measured tardiness in the morning, lost time at the lunch hour, and early departures in the afternoon. The model showed that if the company moved to the suburbs, production would increase 12% with the same size labor force.

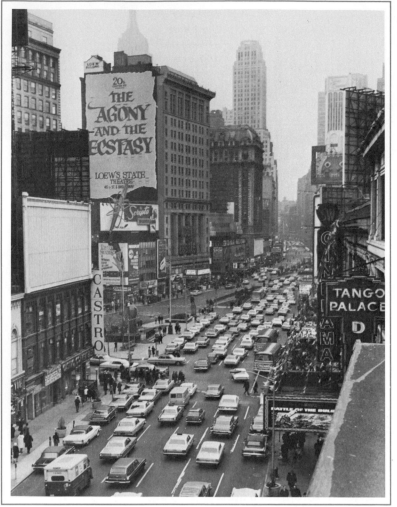

Fig. 4–12.
Broadway as it
approaches Times Square
in New York City. The movie title
is an appropriate commentary.
(*The Bettmann Archive*)

Even if all employees were given free lunches, profits would increase 4% just from the greater production.

Studies based on such models reveal the importance of adequate transportation in the life of any town or city. They also are a commentary on the way business is carried out in a city. When commuting trains, buses, and subways so often bring the businessmen to work an hour late, one has to wonder whether most organizations don't employ more men than they really need.

4 | MODELS FOR RESOURCE MANAGEMENT

Management of resources is a fertile area for the use of models and decision making. In this section, we illustrate this by looking at the history of the buffalo in the western United States. Then we show how this resource could have been managed with some simple models and logical decisions.

The ideas which we present are currently being used in many different ways. In the northwestern United States, the salmon population is being modeled in detail in order to determine how to regulate salmon fishing and changes in water to ensure that the salmon supply will continue in the future. France has several programs for sea "gardening," growing fish and shellfish for food. The invasion of the Great Lakes by lampreys (eels) and the resulting changes have been under study for more than two decades, with a variety of decisions made to restore the fishing industry and sport.

The buffalo provide an interesting example of what is apt to happen with no national policy for resource management—no model, no logical decision.

A Model for 1830

In 1830, there were 40 million buffalo roaming the western United States. In terms of weight (at 1000 pounds each), there were 40 billion pounds of buffalo compared to only 24 billion pounds of human beings today in the entire United States. The buffalo dominated the western plains to an extent unknown for any other animal in history, including human beings.

In 1830, the railroad arrived and the rapid westward expansion of the United States began. By 1887, there were only 200 buffalo left. In this slaughter, animals were often killed for only the tongues and hides. An average of only 20 pounds of meat (of a possible 500) per buffalo was eaten. The peak was reached in 1872 when national heroes like "Buffalo Bill" Cody led the killing of more than seven million.

In less than sixty years, the lack of any sort of policy led to the destruction of what could have been a major source of meat for today's population of this country. A single buffalo could provide the entire meat supply for at least five people for a year.

Had the nation been aware of the potential problem in 1830, a model could have been constructed. In order to decide how many buffalo can be killed each year, we first must know how the population rises or falls from natural causes. Recent studies of buffalo have indicated the following facts:

1. Buffalo reach maturity at age 2.
2. 90% of the females age 2 or older have one calf a year.
3. 53% of the calves are male, 47% female.
4. 30% of the calves live for two years, or to maturity (infant mortality is high for most animals).
5. 10% of the mature beasts die each year (from tuberculosis, drowning, predation, and so on).

Fig. 4–13.
"Buffalo Bill"
Cody. The picture is
from a poster announcing
the appearance of
"Buffalo Bill"
sometime during
the 1890's. (The
Bettmann
Archive)

From these observations, we can construct a model in mathematical form to show the number of mature females alive at the beginning of each year.* For example, let's look first at year seven. How many females are there at the beginning of year seven? Of course, this can be any year picked at random.

First, the answer depends on the number there were at the beginning of year six. Ten percent of those have died. Some other fraction have been *harvested* (killed for meat and hides). Let us call this fraction k. Then, of those who started year six, $(0.9 - k)$ as many start year seven.

This is not quite the whole story. During year six, some female buffalo have reached maturity (two years old). For every 100 mature females at the beginning of year five, 90 calves are born; 47% of these, or 42.3, are female; 30% of these, or 12.69, live to be two years old. Thus, 0.1269 of the female population at the start of year five become mature females at the start of year seven.

The same general sort of model is used for males. The total population is, of course, the sum. To simplify the discussion, we consider only the females. (The male calves born depend on the number of females, so it is easiest to find the number of females first.)

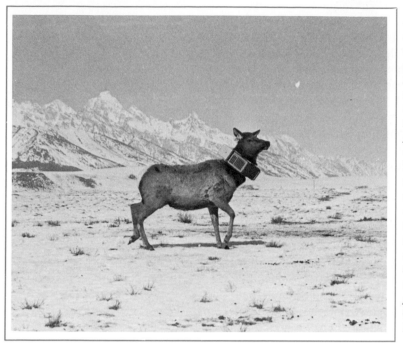

Fig. 4–14.
*A wild elk
outfitted with
a radio transmitter
so his movements can
be monitored from a
satellite. Twice a day,
a radio signal is transmitted
from the satellite which causes the
elk's transmitter to send out a signal.
This signal is received at the
satellite and its origin
measured within a mile.
In this way, we can learn
in more detail about the
life habits of animals
and construct more
accurate models to
ensure the species is
preserved.
(Wide World
Photos)*

Therefore, the number at year seven is $(0.9 - k)$ times the number at year six plus 0.1269 times the number at year five. If we call F_7 the number of females at the start of year seven, we can summarize this model very simply by an equation

$$F_7 = (0.9 - k) F_6 + 0.1269 F_5$$

We can change the 7, 6, 5 to any three consecutive years. All this equation says is that the mature population depends on the size a year earlier and also two years earlier. The first term arises from the total death rate, the second term from the births two years ago.

A Decision Policy

Once the model is known, we can choose k to control the population. We might want the population to increase, stay constant, or decrease.

As an example, if the decision in 1830 had been to keep the population constant at 20 million mature females, how should k be chosen? Now in the preceding equation, we want F_7, F_6, and F_5 to be the same (20 million). Then we can cancel these F's or divide each term by F. This leaves an equation

$$1 = (0.9 - k) + 0.1269$$

which states that k should be 0.0269. This means we can harvest 2.69% of the mature females each year. The population will not

Fig. 4–15.
*Seal of the U.S.
Department of the
Interior. It is ironic that
the animal we slaughtered
so enthusiastically is the
symbol of a government
department and even
appears on our coins.*

change. Since 0.0269 times 20 million is 538,000, we can harvest for food and hides 538,000 mature females each year. (Even more males can be harvested, since more males are born.)

If both males and females are counted, we could have provided enough meat for 6 million people a year since 1830 and still had a buffalo population of 40 million. All that was missing was an intelligent decision policy.

If we were actually managing the buffalo population, we would have to measure the death and birth rates each year. If there were an epidemic, we might want to decrease the harvest until things were back to normal.

We could also construct a more detailed model. Birth rates depend on the age of the mature females, the weather, and other factors. Infant mortality varies with location (the number of predators, the food supply, and so forth).

The interesting feature of the example is the obvious benefit from a very simple model. Even if the numbers in the model are slightly in error, the decision (or policy) ensures that the population changes will be slow. If after several years we find the population is decreasing, we can reduce the harvest slightly. Over a period of years, we can reach a policy which gives a stable population. We can continually improve our model and our policy with experience.

5 | A POPULATION MODEL

While we worry about saving the buffalo or the falcon from extinction, there is much greater concern today about trends in the population of human beings. A model of the world's population is interesting because it represents a dynamic situation (one in which events change with time). Furthermore, the derivation of the model illustrates how we often need to refine or improve our model after it is initially determined.

It has been estimated that since the appearance of man on the earth, a total of 15 billion human beings have existed. With a world population of nearly 3 billion today, 20% of all the people who have ever lived are alive today. Our population is growing at an explosive rate.

Demography, or the study of population, is of increasing concern to economists, ecologists, sociologists, political scientists, engineers, and many others who must understand the present and plan for the future. Models of population change are exceedingly important to such study. They make possible analysis and prediction which can lead to more effective planning for the many goods and services that people need.

We wish to obtain a simple model which would let us estimate the world population at some future date. The present average rate of population increase for the entire world is estimated to be close to 2% per year, and we assume in this section that this rate of increase does not change. At the start of 1960, population was approximately 3 billion (that is, 3 followed by nine zeros, or 3×10^9).

If the rate of increase is 2% per year, by the start of 1961 the increase is 0.02(3,000,000,000), or 60,000,000 people, to make a total of 3,060,000,000. We can then calculate the increase for the next year and for all succeeding years. The results are shown in Table 4–2.

Year	Population at start of year	Increase	Population at end of year
1960	3,000,000,000	60,000,000	3,060,000,000
1961	3,060,000,000	61,200,000	3,121,200,000
1962	3,121,200,000	62,424,000	3,183,624,000
1963	3,183,624,000	63,672,480	3,247,296,480
1964	3,247,296,480	64,945,930	3,312,242,410
1965	3,312,242,410	66,244,848	3,378,487,258
1966	3,378,487,258	67,569,745	3,446,057,003
1967	3,446,057,003	68,921,140	3,514,978,143
1968	3,514,978,143	70,299,562	3,585,277,705
1969	3,585,277,705	71,705,554	3,656,983,259

*Table 4–2.
Estimated world
population, 1960 to 1969.
These data can be compared
with more recent estimates: In
July 1967, the Population Reference
Bureau used United Nations and
other statistics to estimate
that, in the summer of
1966, world population
was 3.34 billion, an
increase in one year
of 65 million.*

The table shows that the increase is *greater* each year. The growth is always 2% of the population at the beginning of the year. As the population rises, the growth also increases. In fact, if we continued Table 4–2, we would find a population of 6 billion by 1995. That is, the population will double in about 35 years.

Carrying the calculation further, we would notice that this doubling occurs *every* 35 years. This would be true for any population number we start with, as long as the rate of increase is 2% per year. If the rate of increase were 3%, the doubling would occur in 23.5 years.*

Are these numbers in our table really accurate? From the entry in Table 4–2 for the population at the beginning of 1966, the model predicts exactly 3,378,487,258 people, a precise value. We have, however, ignored the fact that the numerical values with which we started were only approximations: The 3 billion initial population was a rounded number, and the 2% was an estimated average growth rate. If the initial population was *exactly* 3,000,-000,000 and the rate of increase was exactly 2.000,000,000%, then we should obtain ten meaningful digits in our answer. But since precision was lacking in our measurement of both the starting population and the rate of increase, the results can have only a limited number of significant figures. We must, therefore, be content to use rounded numbers. The extent of precision when two

* This is the same growth as compound interest gives. A rough rule is that the time to double is 72 divided by the rate of increase or interest in %. Thus, 5% interest doubles the money in $\frac{72}{5}$ or a little over 14 years.

numbers are multiplied is restricted by the number with the smaller precision. If in this case it is the 2% figure, and we assume that we are certain of its value to three significant figures (2.00%), then the rounded number having acceptable accuracy is not 3.378487258 billion, but 3.38 billion.

If we continue our example with a 2% rate of increase, we find that in the year 2060, just one hundred years from our starting date, the population will be nearly 22 billion. By the year 2160, it will reach an enormous 157 billion! With a doubling of population in 35 years, the growth after two centuries results in a population which is more than fifty times the original population.

Plots of Population Growth

We have already seen that a plot or graph is much easier to understand than a table of numbers. We now construct such a plot. For this, we use the predicted population at the beginning of each decade from 1961 to 2060 (Table 4–3).

Decade	Population (in billions)
1961–1970	3.06
1971–1980	3.72
1981–1990	4.55
1991–2000	5.55
2001–2010	6.77
2011–2020	8.20
2021–2030	10.06
2031–2040	12.2
2041–2050	15.0
2051–2060	18.3

Table 4–3.
Estimated
world population at start of
each decade, 1961–2060.

As a matter of convenience, we have used the population for the first year of the decade, although the actual number continually grows. In the United States, where a census is made every tenth year, the count obtained is often considered to be the legal population until the next census is completed, even though the Census Bureau issues an annual estimate of the current number of our people. These values are plotted as a bar graph in Fig. 4–16. The height of each vertical column is proportional to the population at the start of that decade as given by the table. Not only do the heights of the bars go up in each ten-year period, but the steps become increasingly larger.

The bar graph is one way of plotting the population growth. At the beginning of each decade (the start of 1961, 1971, and so forth), the figure is an accurate estimate or prediction. Then for the next ten years (Fig. 4–16), we show the population as constant

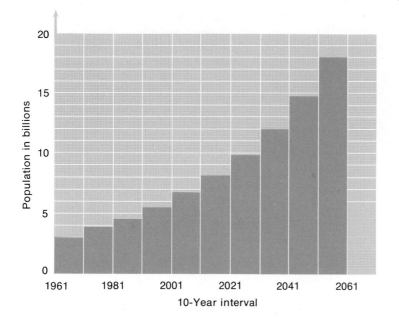

*Fig. 4–16.
Estimated
growth of world
population.*

until the start of the next decade. Actually, we know that the population tends to increase fairly smoothly year by year, month by month, and day by day.

This steady growth in population can be portrayed if we

1. Plot the points corresponding to the start of each decade (the points shown in Fig. 4–17).
2. Draw a smooth curve through these points, as shown in the figure. This smooth curve gives only an approximate model of the real situation, since the population growth does fluctuate from year to year due to famine, epidemics, wars, and so forth. The curve does represent a prediction of world population if we assume:

   a starting figure of 3,060,000,000 in 1961 (or 3.06 billion at the start of 1961) and

   a growth of 2% per year.

Now the extremely fast growth of population is clear. Even though the percentage increase remains constant at 2% per year, the larger increases each year produce a curve which becomes steeper and steeper. This curve is different from those we found in our previous models. The previous plots were linear. The population curve of Fig. 4–17, however, is not a straight line.

Furthermore, this is a particular kind of curve. Each new value of the variable is obtained by adding a constant percentage of the previous value to that value. We have a growth that is proportional to the accumulated size: "the bigger it gets, the faster it grows."

Fig. 4–17.
Fitting a smooth
average line to the population
growth graph.

A snowballing relationship of this type is called *exponential*. The curve of Fig. 4–17 is known as an exponential curve. This very important curve represents a model which is encountered frequently, and we find other examples later in this chapter.

Population Plots Over Longer Time Scales

We now consider another plot of population increase, from 1700 to 2165. This is shown in Fig. 4–18. The eye tends to follow the curve upward to the right; but it is also important to note that the graph drops as we look to the left, or as we go backward in time. In fact, prior to 1800 the height of the curve on the scale of this graph is so small that it is difficult to measure its value. This reflects the fact that a "population explosion" has occurred: The population of the earth in the past was extremely small compared to the present population. The curve makes more reasonable the earlier statement that approximately 20% of all the people who have ever lived are alive today. From a larger scale copy of the curve we could find the even more striking fact that the population *increase* from 1940 to 1963 (just 23 years) was greater than the total population of the world in 1800!

Since we have extended our look backward in time, it is appropriate also to look further into the future. We might attempt to look ahead to the year 2700, a period only slightly more than 700 years

Fig. 4–18.
The growth of
world population from
1700 to 2165. Prior to the
present time the curve is based
on historical fact. Later
values are predicted
from a model.

from now. This represents about the same time difference as that between the present and the time of Marco Polo. The graphical results of the computations are shown in Fig. 4–19.

Can this really be expected? The curve shoots up at a fantastic rate. The vertical scale on the left is much larger than in the preceding figure—so much so that the steeply rising curve to the year 2165 (Fig. 4–18) cannot even be seen. Our new exponential curve has reached such proportions by the year 2700 that, if we tried to plot it on the scale of Fig. 4–18, we would need a sheet of paper 27 thousand times as high, or 11 thousand feet (more than two miles) high instead of five inches.

What does this curve of Fig. 4–19 tell us? By the year 2510 we should expect to have a world population of nearly 200,000 billion people, by 2635 about 1,800,000 billion people. Thirty-five years after that it will have doubled to approximately 3,600,000 billion. In the year 2692, the model predicts a 5,450,000 billion population.

How large is 5,450,000 billion? We can express it in many ways. The number when written completely would appear as:

$$5,450,000,000,000,000,000$$

It may be written as 5.45×10^{15}, but this form does not give one a good "feeling" for the enormous size.

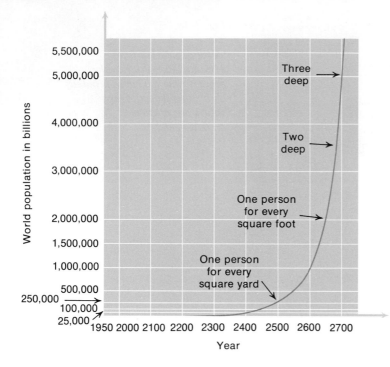

Fig. 4–19.
Modeling prediction
of world population to the
year 2700.

Here is one picture that helps to visualize the magnitude of the number. There are roughly 31 million seconds in each year. If we counted one thousand persons per *second*, it would take us 176,000 years to complete the census.

There is yet another way to grasp the significance of this estimate of 5,450,000 billion population. Let us ask where these people will be: How much room will they have? The surface of the earth contains approximately 1,860,000 billion square feet. About 80% of this area is covered by water, but let us suppose that all of the surface were land. We can calculate that in the year 2510 when the population is 200,000 billion, there will be 9.3 square feet per person, or about one person per square yard *all over the earth*. Worse yet, in 2635 each person will only have one square foot on which to stand, and in 2670 if they insist on retaining that much real estate, they will be standing on each other's shoulders *two deep*. And only 22 years later they will be *three deep*. Now, if we do not assume that these people can tread water but instead must occupy the land area only ($\frac{1}{5}$ of the total area), in 2692 we should expect to see totem poles 15 persons high on every square foot!

6 | EXPONENTIAL GROWTH

The model of the last section obviously is not useful for predicting population too far into the future. The model is based on a 2% increase every year. Once the land becomes too crowded, this growth rate certainly will fall. The population will level off. In the next section, we will consider how to change our model to include this fact.

The model is useful, however, to predict what will happen in the next thirty years. If the present growth rate is 2% per year, we can guess that this will not change very much in the near future. Regardless of the international efforts to control births, any real change requires educating masses of people. Even if reasonable laws were passed to limit the number of children a family could have, years would pass before the laws could be enforced effectively.

Furthermore, there is a time lag involved in any program of population control. This is a factor in the model which is often overlooked when people write or speak about the population problem. To reduce the number of babies born each year, we need to reduce the number of women of child-bearing age. But the number of women 20 to 40 years old depends on the number of babies born more than twenty years ago.

Consequently, even if the number of babies born this year were reduced, it would be more than twenty years before the effects were really noticeable. In other words, the population growth rate is fairly well determined already for the period from now until the year 2000. The real effects of any population-control program would not show up until about the year 2000.

Such a time lag is found in problems where we are trying to control the environment. For example, even if all new cars built from today on were made to have no lead compounds in the exhaust gases, it would be several years before there was a noticeable change in the lead in the air simply because of all the cars already on the road. Ten years would pass before almost all of the cars on the road were "clean."

Because of this time lag, the model based on 2% growth per year gives a reasonable prediction of population for the next thirty years. (We assume there will be no world war, great epidemic, or mass starvation.) In this section, we want to look in more detail at models which have such a constant growth rate.

A Plot with Constant Growth Rate

A signal (or a quantity such as population) with a constant growth rate is called an *exponential*. This simply means that the *percentage change* in each year (or hour or second) is the same. Each year the population increases by 2%.

When we try to plot population, we run into the problem shown in Fig. 4–20. A plot which shows what is beginning around the year 2000 is useless before 1800 (where the curve is almost zero on our vertical scale). Also, the curve starts to rise so rapidly after 2100 that it quickly goes off the paper. The trouble is clear. Our vertical scale only allows us to plot values from perhaps 2 billion

Fig. 4–20.
The growth of
world population from
1700 to 2165.

to 200 billion. Over the years of interest, the population varies over a much wider range than this.

When we have a constant growth rate, we can simplify the plotting if we use a different vertical scale. The population is 3 billion in 1960 and we are interested in the years from 1760 to 2160. How do we plot?

We first take a sheet of paper and mark the years off in the regular way. 1960 is in the middle, and every 50 years are shown from 1760 to 2160. We know the population in 1960 is 3 billion, so the middle of the vertical axis is set at 3.

In the last section we saw that the population doubled every 35 years (with a 2% increase per year). Hence, we have the points

Year	Population
1995	6
2030	12
2065	24
2100	48
2135	96

Along the vertical scale, we show a doubling every equal space (Fig. 4–22). In other words, the distance from 3 to 6 is the same as from 6 to 12. Once the vertical axis is labeled in this "odd" way, we can mark the above points.

The usefulness of this plot is now obvious. *The points are on a straight line.* The population doubles every 35 years. If the vertical axis is marked to show a doubling corresponding to a constant distance up, the curve moves up this same amount every time it moves across by 35 years. This is just the property of a straight line. If we draw the rest of the curve back to 1760, we obtain the complete graph of Fig. 4–23.

Once the horizontal and vertical scales are labeled, we need only two points to determine the straight line. For example, the 1960 population is 3, and the 1995 is 6. We can just draw a straight line through these two points.

Fig. 4–21. Years marked off for the population plot.

Fig. 4–22. Points into the future shown with an unusual vertical scale.

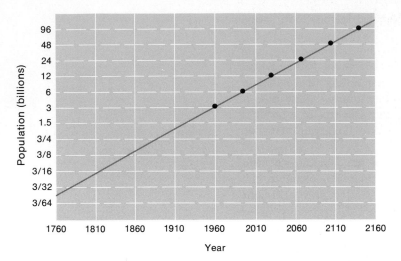

Fig. 4–23.
Plot of population
growth from 1760 to 2160 with
increase of 2% each year.

An Economic Model

The plot of Fig. 4–23 is useful whenever there is a constant rate of growth. The economic plots of the United States gross national product, for instance, show an average increase of 5% each year. If the total is 1 trillion dollars in 1970, the future can be predicted from Fig. 4–24. We construct this plot as follows.

The years of interest are 1960 to 1990, so we label the horizontal axis. The 5% rise each year means we double every $\frac{72}{5}$ or about every 14 years. We now have two points (1 trillion in 1970, 2 trillion in 1984). The straight line can then be drawn.

The graph shows that in 1990 the gross national product will be almost 3 trillion dollars. Since the United States population in 1990 is expected to be about 300 million, the average per person income should be 3 trillion/300 million or $10,000. (In 1970 it was about $5,000.)

One word of caution should be given here. To make Fig. 4–24, we have used an odd vertical scale (Fig. 4–25). The distance from 1 to 2 is the same as from 2 to 4. Each time we move this distance, the quantity being measured doubles. When we asked in Fig. 4–24 what the gross national product would be in 1990, we had to estimate a value between 2 and 4 (point A in Fig. 4–24).

To make this estimate, we must recognize in Fig. 4–25 that if a is halfway from 1 to 2, a does not correspond to 1.5 trillion. If it did, going from 1 to a would mean multiplying by 1.5. Going from a to 2 would mean multiplying by $\frac{2}{1.5}$ or 1.33. But we said equal distances on the vertical scale correspond to equal multiplying factors. If this is to be true, the point a corresponds to $\sqrt{2}$ trillions, or 1.4 trillions; b midway between 2 and 4 is $2\sqrt{2}$ trillions or 2.8 (not 3).

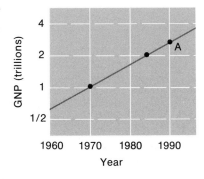

Fig. 4–24. Expected gross national product of the United States.

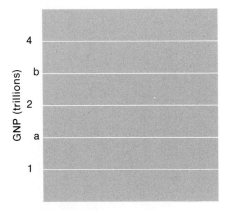

Fig. 4–25. Vertical scale for an exponential plot (a is 1.4, b is 2.8).

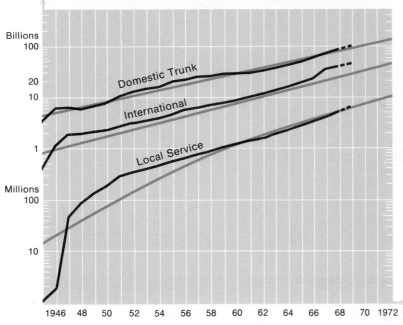

GROWTH TRENDS IN U.S. AIRLINE PASSENGER TRAFFIC

Billions
100

20
10

Domestic Trunk

International

1

Local Service

Millions
100

10

1946 48 50 52 54 56 58 60 62 64 66 68 70 1972

Fig. 4–26.
Another example of
a plot of exponential
growth. The way in which
the actual data follow the
straight lines promises that the
changes in the next few years can
be predicted with reasonable confidence.

These particular curves are used by
airport planners, airlines, travel agencies,
government agencies, and companies
building equipment for the air-travel
industry.

The straight line for international
travel shows that the travel doubles
every 5 years (1962 to 1967).
Thus, the annual increase is
72/5 or 14%. (Aviation
Week and Space
Technology)

The Solid-Waste Problem

If the United States is considered as one vast system (containing as elements the manufacturing plants, the transportation vehicles, the people, and the natural and man-made devices), one of the important signals which can be measured is the solid waste. Such solid-waste material includes all the material which we throw away: the 6 million cars which are scrapped every year, the appliances discarded, the refuse from construction and demolition of buildings, and the garbage. Our highly advanced technology and the associated high standard of living lead to a national problem of increasingly serious magnitude. How can we dispose of this solid waste economically and without dangerously fouling the environment? If we take a longer range viewpoint, how can we recycle this trash? In other words, how can we conserve our natural resources by using again the materials we are throwing away?

The magnitude of the problem is vividly portrayed in Fig. 4–27,* which shows the quantity of solid waste produced per year in the United States since 1920. The significance of this particular signal is perhaps clearer if we note that in 1965 nearly five pounds were produced each day for each person in the country. Furthermore, the rate of increase is appreciably greater than the rate of population increase. (Indeed, the United States is by far the most efficient producer of rubbish in the history of civilization. With less than 10% of the world's population, we generate well over half of its rubbish.)

Figures 4–27 and 4–28 are taken from the report, "A Strategy for a Livable Environment," published for the U.S. Department of Health, Education, and Welfare, June, 1967.

Fig. 4–27. United States solid waste.

The importance and urgency of the problem derive from two principal factors:

1. In most cities, available land for dumps is being rapidly exhausted. At the same time, the nature of the solid waste is changing. A few decades ago, the rubbish was primarily garbage and ashes. Today it includes vast quantities of metals, plastics (e.g., non-returnable containers), and other new products, many of which cannot be economically burned without contributing to air pollution.

2. We know very little about the effects of environmental pollution on our physical and mental health. To what extent are polluted air or mounds of junked automobiles responsible for the increases observed in mental illness, in urban unrest, and in such physical illnesses as lung cancer? Even data such as shown in Fig. 4–28 are not easily interpreted. Individuals living in the city may smoke more heavily, may lead lives under greater nervous tension, and so forth. The problem of evaluating the importance of environmental pollution is further complicated by the realization that major effects on the balances of nature and the characteristics of man are unlikely to become evident for a generation or more, when it may well be too late to reverse the established trends. (The problem of the time lag enters again.)

Whether we are worried about a single city or a whole region, the intelligent planning of new methods for disposing of waste requires that we predict how much waste will be produced in the future. For example, if we decide to build a new incinerator to

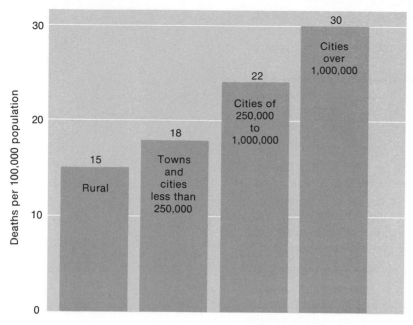

Fig. 4–28.
*Annual deaths from lung
cancer per 100,000 population
as a function of the size of the
community in which the people live.*

Fig. 4–29.
*In a large city-apartment
complex, trash disposal is a
major problem. Incinerators
pollute the air. To remove pollution
economically, the trash must be burned
in an expensive central incinerator
(equipped with stack emission controls)
or collected and hauled away.*

    *The picture shows a pneumatic
tube system for collecting trash.
Tenants place their trash in a
chute. It is drawn to a central
point by an air stream moving
at 60 miles an hour. The
system is designed to put
an end to odors, littered
basements, and cans of
exposed garbage.
(Envirogenics Co.)*

burn the rubbish, six years may be required until the incinerator is operating. Thus, if a decision is made in 1970, the incinerator which is built should be designed for the quantity and type of rubbish which will appear after 1976.

In this discussion, we consider only the quantity of solid waste produced. Figure 4–27 shows the past history of the system over the years 1920 to 1965. We wish to use the data to predict the signal at least a few years in advance. This need for prediction arises for two reasons:

1. Data are usually available only some time after they are valid (in problems of this broad a nature, a year or two may be required). Thus, the curve of Fig. 4–27 runs only to 1965, even though it was published nearly two years later.
2. Design and construction of the facility require several years, so that the system is truly being built for the future.

Before we try to use the data of Fig. 4–27 to predict solid waste in the future, we notice that the quantity is increasing at a constant rate. Every 12 years, the amount increases by 50%. This is exactly the property of an exponential. Consequently, we can change the plot of Fig. 4–27 to the exponential form.

In 1950, there are 100 million tons generated. In 1962, the figure is 150 million tons. We label the time axis from 1920 to 2000. The vertical axis includes 100, and then successive lines 1.5 times as much (Fig. 4–30).

Once the straight line is drawn, we can predict the quantity probably generated at any future date. For example, by the year 2000 we can expect almost 500 million tons per year. With 300 million people, this is about 1.6 tons or 3200 pounds per person per year. (Almost 10 pounds a day!)

One proposal has been to build an artificial mountain of rubbish from New Jersey to California. When it is completed, it could be covered with grass and landscaped so it would provide a recreation area for future generations. The proposal does not sug-

Fig. 4–30.
Solid waste generated with vertical scale chosen to show 50% every 12 years.

Fig. 4–31.
One of today's
pressing problems is
abandoned cars. The number
of abandoned cars grows
exponentially. Attempts
to deal with the
problem have been
essentially
unsuccessful.
(Photo by Richard O'Leary,
St. Louis, Mo.)

gest how to compensate the people who happen to live nearby and have to watch it grow.

More attractive proposals are to raise the cost of goods so that the manufacturer has to reclaim them. Automobiles are a prime target. In 1969, more than 50,000 cars were simply abandoned on the streets of New York City. Any person driving through the United States soon encounters automobile junk yards steadily growing in size and ugliness.

The Exponential As a Planning Device

In this section, we have seen that exponentials can be plotted as straight lines if the vertical scale is marked as follows. Each unit distance up corresponds to multiplying the quantity by the same factor.

Because exponentials occur so often, this form of plot is particularly important. In using such plots for prediction of the future, it is important to recognize that we usually do not need great accuracy. Whether the solid waste generated in the United States in the year 2000 is 500 million tons or 450 million tons is not very important. Regardless of what the number will actually be, we need to start planning immediately if we are not to be swamped by rubbish. Either we have to stop its growth or we must find better ways to recycle it or dispose of it.

Box 4–1

EQUATIONS FOR EXPONENTIALS

For those who like to live by mathematics, we can write an equation for an exponential growth curve. The solid-waste generation data are used as an example.

The general equation for an exponential is

$$y = A \, r^{t/T}$$

Here y is the quantity we are measuring (the solid waste generated per year in millions of tons). r is the factor by which y is multiplied in T years. In our case, r is 1.5 and T is 12 years. t is the time (in years) measured from any reference point. If we select the 100 value in 1950 as the reference, t is measured from 1950. (Thus, a t of 7 represents 1957, a t of -21 represents 1929.) Finally, A is the quantity at the time t is zero. (In our case, A is 100 since that is the solid waste in 1950.)

Thus, the equation for United States solid-waste generation is

$$y = 100 \, (1.5)^{t/12}$$

where t is measured from 1950. In 1998, t is 48 and

$$y = 100 \, (1.5)^4 = 506$$

The equation predicts 506 million tons generated in 1998.

The equation gives exactly the same information as our straight-line graph.

This type of prediction is called *extrapolation*. We extrapolate (or extend) what has happened in the past into the future. We assume that the growth rate will not change. We should recognize that such extrapolation is not necessarily valid. New changes may occur in our way of living which may increase or decrease the rate of population growth or solid-waste generation. (See Box 4–1.)

7 | AN IMPROVED POPULATION MODEL

Obviously the population model with 2% growth is incorrect, at least after some time in the future. Quite clearly there are some limiting factors which prevent such a population increase. Actually our model is too simple because we did not take into account several important factors that tend to limit our predictions.

To learn more about these factors, it is helpful to examine functional models of the world population. Such models are easy to find in a biology laboratory. Any small organism that reproduces rapidly will do. Fruit flies, yeast, and bacteria are commonly used examples. Here we describe a population model using yeast.

First the experimenter must prepare a food supply, a "nutrient medium." For many yeast species, this may be simply a weak sugar syrup slightly modified by addition of other substances. Then there is need for a jar in which to keep the yeast as they gorge themselves, and for Adam and Eve, so to speak: the syrup must be inoculated with a few yeast cells to start with. The temperature should be kept constant, and the medium should be gently but constantly stirred.

It is hardly possible to take a census of yeast cells as one does of people. Instead, a sampling technique is used. Knowing the starting volume of his experiment, the investigator can withdraw a definite, very small percentage of it and count the yeast cells in that. Since the solution of food has been stirred, he can safely assume that his sample is typical, and that he can simply multiply by the proper factor to learn the total population. Such models as this one are particularly convenient because they take up very little space; moreover, it is easy to try different circumstances ("to vary the parameters" as the professional puts it). It becomes possible to answer such questions as these: What is the rate of increase of population when the experiment begins? Does this rate remain constant as the population becomes larger? Does it matter whether the available space for the organisms remains constant or is made to increase as the population grows? Yeast cells produce an alcohol (there are many kinds of alcohol) from the sugar they consume; what is the effect of leaving the alcohol to accumulate in the nutrient medium? Of removing all but a constant fraction? Of removing all of it as it forms? (Removal can be rather easily accomplished by continually pumping fresh nutrient medium in and at the same time allowing the used medium to trickle out through a filter.)

Which of these possible experiments cast light on our graphical model of world population? First, we know that the entire land surface of the earth is not inhabited but that it is not unlimited. (There is room for population to increase but the space will be used up someday.) This is modeled in the yeast case by using bigger jars (and more medium) up to a certain point, but then no more. Second, we know that food production can be increased for human beings but not without limit. We can supply more sugar to the yeast on a schedule that we think is comparable to the future history of the world; even better, we can try many schedules. In short, we can test our model and thus refine it, by comparing it with what we already know about the course of development of the human population.

Now it turns out that such experiments as those described are practically always alike in one feature. Growth is roughly exponential at first; the *rate* of increase is not necessarily constant, but the population curve is closely similar to that of Fig. 4–32. If the

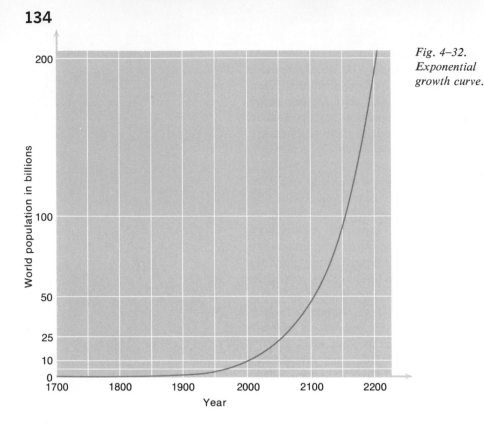

Fig. 4–32. Exponential growth curve.

experiment lasts long enough, however, the rate of growth sooner or later begins to decrease and in time reaches zero. The curve stops its exponential growth and tends to level out, as suggested in Fig. 4–33. Because this curve has a kind of S-shape, it is known as a *sigmoid* (from the Greek word *sigma* for the letter S).

The basic reason for the change of shape shown is overcrowding. Without unlimited space in which to grow and unlimited food to support life, the individual yeast cell has neither room nor food to allow it to reach normal size. No doubt there are other reasons, but they are less important.

In the case of small animals and insects tested in a laboratory or carefully controlled environment, the same leveling off of population has been observed. The rate of growth tends to decrease as the animals become badly overcrowded because of both physical and mental deterioration resulting from inadequate supplies of food, air, and water, excessive nervous tension associated with inability to move freely, and so forth. In actual experiments with rats, once severe overcrowding occurs, mothers reject their young, adults kill females, and reproduction falls rapidly. Whether such experiments have any meaning for human beings is not at all clear.

Certainly nature ultimately limits population growth, although typically with a serious deterioration of the species as suggested in the preceding paragraph. One would assume that man will limit his population growth significantly before serious damage has been done to his general physical and mental condition. Indeed, the focus of the recent worldwide emphasis on population control is to limit the population to levels at which adequate food resources and land are available to ensure happy and healthy individuals.

Perhaps the first man to recognize the sigmoid character of population growth (although he did not express it in this way) was Thomas Robert Malthus (1766–1834). He was an Englishman who wrote a gloomy essay pointing out that a time must come when population will outrun food. Then the growth of population would be stopped, he believed, by widespread epidemics, or starvation, or war, or some combination of these. Instead of a truly sigmoid curve, however, his curve would probably actually turn downward. He predicted that the end of the growth period would come during the nineteenth century. That it did not was the result of the discovery of chemical fertilizers. With these an acre of ground brings forth several times as much food as was possible in Malthus' day. We can see, however, that some limit, at some time, must be reached in the number of pounds of food that can be won from an acre of ground or of sea. The supplies of potash and phosphate easily recovered must someday disappear, so the cost of fertilizers must rise. The third major fertilizer element, nitrogen, is available without foreseeable limit from the air. The current research in development of food from the sea promises to provide the possibility of feeding adequately a great number of people, but not an unlimited number.

A Model with Decreasing Growth

Even if we agree that the population model of the preceding section (with 2% annual growth rate) is impossible, we can still use this exponential growth curve to predict moderately into the future. The sigmoid characteristic of Fig. 4–33 does behave as an exponential in the early portion. Recent population data indicate that we are indeed still in this part of the curve.

If we desire to peer further into the future, we need a model which shows the eventual slowing of growth. To obtain such a model, we need to assume that the growth rate will eventually decrease as the population increases.

In the population model, we worked with a growth rate of 2% a year. The population in any year is 1.02 times the population a year earlier. In a stable situation, with the population constant, the value in any year is just 1.00 times the value the preceding year.

Thus, to obtain a sigmoid curve, we need to reduce the 1.02 to 1.00 as the population increases. Exactly how do we put in this change? When does the factor change from 1.02 to 1.01, for example? Then when from 1.01 to 1.00? Where will the population start to level off?

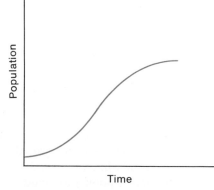

Fig. 4–33. *Population growth in laboratory studies.*

Box 4–2

MATHEMATICS OF A SIGMOID MODEL

The man who relishes mathematical language can derive a model for the sigmoid characteristic as follows.

Exponential growth In the exponential case, the population in any year is a constant (1.02 in our case) times the value a year earlier. If P_{n+1} is the population in the year $(n+1)$, P_n in the year (n), then

$$P_{n+1} = r\, P_n$$

Thus, if P_{1960} is 3 billion and r is 1.02,

$$P_{1961} = 1.02 \times 3 = 3.06 \text{ billion}$$

Sigmoid behavior To obtain the leveling off, the factor r must decrease as population increases. This change occurs if we choose

$$r = r_0 - c\, P_n$$

r_0 is 1.02. This value is reduced by $c\, P_n$ where c is a constant. As the population P_n increases, the r decreases. Then

$$P_{n+1} = (r_0 - c\, P_n)P_n$$

Example If r_0 is 1.029 and c is 0.003, and we measure population in billions, we obtain

$$r_0 - c\, P_n = 1.029 - 0.003 \times 3 = 1.02$$

when the population is 3 billion. That is, in 1960 the growth rate is 2%.

As P_n increases, the factor $(r_0 - c\, P_n)$ decreases steadily. This factor becomes one when

$$1.029 - 0.003 \times P_n = 1$$

or

$$P_n = 9.7 \text{ billion}$$

According to this model the population levels off at 9.7 billion. This value depends critically on what we choose for the constant c, which represents the rate at which the growth decreases. Since we do not know what factors will limit population, we really can only guess at c.

Obviously, we can only guess. Today we are still in the region of exponential increase. This growth will eventually be slowed, but we do not know what factors will cause the leveling off.

The particular population model of this section is not especially important; indeed, major political or social changes may make our predictions of future world population look ridiculous to the historians of the year 2000. Our purpose in this section, however, is to introduce the idea of model refinement. We discovered that our 2% model is ridiculous if we try to predict hundreds of years into the future. On the basis of our rather superficial understanding of population growth, we then found an alternative model which at best has the property of leading to a limited ultimate population. (See Box 4–2.) Since we know we cannot predict population accurately too far into the future, our simple models are probably as good as much more complicated ones.

8 | USES OF MODELS

Predictive population models are often used with great success in governmental planning at all levels—town, state, and federal. For example, the design of a transportation system for a region requires that we have reasonably reliable predictions of population distribution in order to assess future transportation needs (for transporting people and the materials which people require).

In such a problem, the complete system model includes population models for hundreds or thousands of separate towns. The complete model is often a mathematical model composed of many equations. Some of these are similar to the equations we have used and some are more complicated. Not only must birth and death rates (net growth) be considered, but the relationships among other factors must be included. There are also influences which make the populations of towns interdependent. If one town becomes unduly crowded, there is a strong tendency for neighboring towns to grow more rapidly. Immigration and emigration rates thus are important considerations for the development of an accurate dynamic model: a model in which the interrelationship of factors changes with time.

Many Models for One System

It is possible for one system to be represented by a number of different models. As in the case of the blind men and the elephant, no one model describes the real thing completely, but separate models of sub-systems are often necessary and useful.

An air-conditioner provides an example of this characteristic. One model can be developed which is based on heat flow: how heat is extracted from a room, how the fluid in the unit changes its temperature as it absorbs heat, and how this heat is then transferred outside of the room. This model must include such factors as expected temperature ranges, characteristics of the refrigeration unit and blower, and intake and outlet duct air flows.

Another model to describe the same system might be a control model which includes the thermostat, the various relays and contacts, and the electrical network which links the electrical parts of the system.

LOW DENSITY
RESIDENTIAL

MIDDLE DENSITY
RESIDENTIAL

HIGH DENSITY
RESIDENTIAL

INDUSTRIAL

COMMERCIAL, CIVIC

RECREATION,
PARKS, SCHOOLS

← N.J. TPKE. 1 MI.

RT. 33

N

*Fig. 4–34.
An initial model for
the planning of a new
town—in this case Twin
Rivers, New Jersey, with a
planned population of 10,000.*

*This first model shows the
general planning to include the land
uses and the highways and lakes. (The
two irregular areas are lakes.) Once this
overall model is adopted, each section
can be planned in detail. When the
town is designed in this way, schools,
parks, and shopping areas can be
located within walking distances
of most residential areas. Streets
can be planned to avoid too many
pedestrian crossings. The goal is
to create a town which avoids
many of the annoyances and
difficulties of existing towns
which have grown without
overall planning. (© 1970
by The New York Times
Company.
Reprinted by
Permission)*

Yet another model of an air-conditioner could be developed
for a study of its mechanical behavior. For example, we may wish
to know how much noise and vibration the equipment will produce
and how to design the air-conditioner to minimize the noise and
vibration. For this objective the model would include a number of
factors such as the characteristics of the moving parts, their
mountings, the location of shock absorbers, and the geometrical
arrangements of the openings, the absorbent surfaces, and the
baffles.

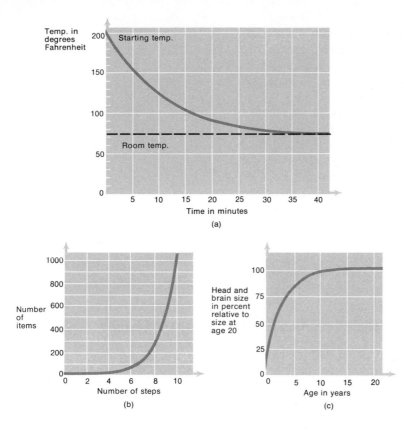

Fig. 4–35.
Three
examples of exponential
systems.

One Model for Many Systems

What is there in common among the way in which a cup of coffee cools, the way in which the numbers of chain letters increase, and the way in which a human head grows? Just as it is possible for one system to be described by several different models, so one model frequently is applicable to many kinds of systems. In Fig. 4–35 there are models of three different processes which one would not ordinarily think of as being similar. In (a) is shown how the temperature of a cup of coffee drops as the coffee cools to room temperature. The initial temperature is just below the boiling point. It drops rapidly at the start, then more and more slowly. After a half hour, the temperature of the coffee has dropped to within a few degrees of room temperature.

The illustration in (b) describes a system in which the production of an item doubles at each step. During the early part of the curve the values are not readily observed because of the scale of the graph axis, but as the number of steps goes from 1 to 2 to 3 to 4, the number of items increases from 2 to 4 to 8 to 16. The increases become larger with each step; for instance in going from 9 to 10 steps the number of items doubles from 512 to 1024. This is a model for the chain-letter process, where an individual writes to two people, each of these writes to two others, and so on. After 20 steps in such a process the number of letters (items) being written is more than one million, and after 30 steps the number becomes greater than one billion.

This rapid growth of an exponential is the basis for the familiar story of the golf game between Mike and Dick. On the first tee, Mike suggested they play for a penny on the first hole and then double the bet each hole. By the eighteenth hole, Dick was so nervous he missed his drive completely, while Mike played with total relaxation. It was only after the match was over that the mathematically slow Mike realized that the bet on the last hole was more than $1300.

Figure 4–35c shows how the size of a human head grows from birth to age twenty. At birth it is a little less than $\frac{1}{4}$ its full size, and it is growing very rapidly. At the age of five the growth begins to slow down appreciably, and at the age of fifteen the head is within a few percent of its ultimate size.

We can see what is common to the cooling of a cup of coffee, the rate of increase in a chain-letter situation, and the growth of the human head. Each displays an exponential rate of change. In each of the processes, as in the population expansion, growth either increases or decreases exponentially.

Our exponential curve thus fits many systems. Such things as the rate at which an automobile coasts to a stop, the growth of plants, and the accumulation of bank interest can also be represented approximately by an exponential curve. All these examples show that one model may represent many different systems.

Another remarkable aspect of modeling is how often we can find mathematical relationships among important quantities. As a final example, violent storms are of four general types: tornadoes, thunderstorms, hurricanes, and cyclones. There is a relationship between the size of the storm and its duration: the larger the storm, the longer it lasts. Indeed, if we call D the diameter of the storm in miles and T the duration in hours, the model is given by the equation:

$$D^3 = 216 \, T^2$$

Once we have found such a model from observation of many storms, we can answer such questions as:

> How long does a cyclone with a diameter of 600 miles last?
> How large is a thunderstorm which lasts one hour?
> If a hurricane has a diameter of 100 miles, what would be the anticipated duration in days?
> A tornado lasts 11.5 minutes; what is the expected size?

The answers are 1000 hours, 6 miles, 2.8 days, and 2 miles.

Such a model is only approximate. Any given storm may deviate from our equation just as any particular seventeen-year-old boy may be 5 feet 2 inches tall and weigh 300 pounds. In spite of such occasional deviations, however, the model yields a picture of typical average relationships among physical quantities.

Determination of Models

Once an idea of the structure and nature of a thing is conceived, it may be expressed in many different ways. We may have different models. Some, as we have seen, are verbal models. A map is a model, a graphical model. Other models are mathematical in which quantitative expressions are used to describe relationships in a precise way. Some models are developed using computers.

An aircraft represents so complicated an aerodynamic problem that a complete mathematical description may be impossible. Therefore, it is usually modeled by constructing a small-scale version of metal or wood for testing in a wind tunnel.

Models are used, not only to describe a set of ideas, but also to evaluate and to predict the behavior of systems before they are built. This procedure can save enormous amounts of time and money. It can avoid expensive failures and permit the best design to be found without the need for construction of many versions of the real thing. Models evolve, and it is customary to go through a process of making successive refinements to find a more suitable model.

For example, in the development of a model for a nerve cell, there is need for successive refinement. A preliminary model is designed, it is tested against the real nerve cell, then the model is modified so that it becomes more realistic in its behavior. In the process of model construction it is essential to alternate back and forth between the real world and the model.

The essential parts of the model-making process are shown in Fig. 4–36. Measurements or observations of the real world are used to develop a model. After a preliminary model is made, measurements made using the model are compared to the behavior

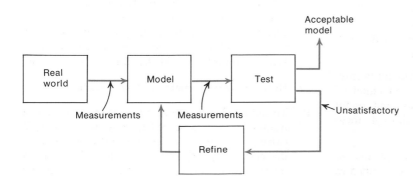

Fig. 4–36.
The model-making
process, shown as a
block diagram.

of the real world. In most cases these tests show that the model is not completely satisfactory, so that it must be refined. This process is repeated until the model is acceptable.

In the modeling of a nerve cell, or the modeling of the growth of a population of people, the real-world measurements are made on a system which already exists. In this case our model-making process is intended to produce a model which accurately matches the real world. In the case where scale-model airplanes or space-craft are modeled, the real-world object may not yet exist, and the box marked "Real World" in Fig. 4–36 theoretically contains the real object which we imagine and wish to achieve, as well as all pertinent facts about the real world (such as the properties of air, characteristics of flight systems which have already been built, and the characteristics of various materials and fuels). The model-building procedures are no different from those already discussed.

Models can be *descriptive*, as in verbal, graphical, or mathematical representations. They can also be *functional* (they "really work"), as in scaled-down airplanes for use in wind tunnels or working replicas of nerve cells. The model includes only those parts of the system which are important for our purposes. The model guides our thinking and suggests how we can improve the system.

Questions for Study and Discussion

1. "No model is ever complete." Would it be helpful if one could in fact construct a complete model? Explain briefly.

2. Discuss the differences between

   a) functional and descriptive models.

   b) dynamic and static models.

   Give an example of each.

3. Suggest two reasons why a mathematical model may be desirable.

4. When a model is first designed, what is the next step which should be taken with it?

5. Figure 4–3 shows the height-weight data for twenty-year-old men as a somewhat scattered cloud of points. Explain why it is reasonable and useful to draw a *particular* straight line through these points.

6. The greatest height shown by the graph line of Fig. 4–4 is about 6′5″. Would you be justified in using the graph line to predict the weight of a candidate for center on the basketball team if he is 6′10″ tall? Why, or why not?

7. From the equation for the relationship between weight and height of twenty-year-old men, find the expected height of a young man who weighs 145 lbs.

8. A forester might use a method similar to that of Section 2 to study the relationship between the height of a tree and the diameter of its trunk. It was pointed out that the height-weight data had been "passed through two strainers (age and sex)." What "strainers" would the forester have to use?

9. How do highway engineers take a traffic count?

10. To use an electric eye as a sensor in a corridor, a beam of light is arranged to shine from one wall (where the light source is) to the other wall (where the electric eye is). Whenever the light beam is interrupted by a person passing, the circuit of the

electric eye operates a counter. Why would such a device give an incorrect count in the situation shown in Fig. 4–8? Can you think of a way to make it count correctly?

11. The text discussion on the uncontrolled harvesting of buffalo is but one example of how man has willfully brought a species of animal to the point of extinction. Describe five other examples of how man has either willfully or unwittingly produced this effect.

12. The movements of the elk pictured in Fig. 4–14 are monitored from a satellite. Describe how this is done and explain what elements of the life habits of an animal can be learned in this way.

13. In the text, several factors which affect the growth rate of a town are listed: birth rate, death rate, nearness of other crowded towns, transportation system. Suggest five other factors which might influence the growth rate of a town. In each case, explain briefly why the factor would be likely to increase or to decrease the growth rate.

14. We have seen that it is possible for one system to have a number of different models which apply to it. Many times an engineer finds it necessary to have models of sub-systems. Suggest three models which could be used to describe a submarine.

15. The poem on the blind men is:

THE BLIND MEN AND THE ELEPHANT

It was six men of Indostan
To learning much inclined,
Who went to see the Elephant
(Though all of them were blind),
That each by observation
Might satisfy his mind.

The First approached the Elephant,
And happening to fall
Against his broad and sturdy side,
At once began to bawl:
"God bless me! but the Elephant
Is very like a wall!"

The Second, feeling of the tusk,
Cried, "Ho! what have we here
So very round and smooth and sharp?
To me 'tis mighty clear
This wonder of an Elephant
Is very like a spear!"

The Third approached the animal,
And happening to take
The squirming trunk within his hands,
Thus boldly up and spake:
"I see," quoth he, "the Elephant
Is very like a snake!"

The Fourth reached out an eager hand,
And felt about the knee.
"What most this wondrous beast is like
Is mighty plain," quoth he;
" 'Tis clear enough the Elephant
Is very like a tree!"

The Fifth who chanced to touch the ear,
Said: "E'en the blindest man
Can tell what this resembles most;
Deny the fact who can,
This marvel of an Elephant
Is very like a fan!"

The Sixth no sooner had begun
About the beast to grope,
Than, seizing on the swinging tail
That fell within his scope,
"I see," quoth he, "the Elephant
Is very like a rope!"

And so these men of Indostan
Disputed loud and long,
Each in his own opinion
Exceeding stiff and strong.
Though each was partly in the right
And all were in the wrong!

John Godfrey Saxe
American Poet 1816–1887

A modern version of the six blind men and the elephant is suggested by the following problem. A printed capital letter of the English alphabet is scanned photoelectrically and the resultant signal is read into a digital computer. Seven subroutines in the digital computer inspect it. The first states that the letter is like a U because it has at least one pocket to hold rain coming from above; the second shows that it is like a K because it has at least one pocket to hold rain from below; the third and fourth find that it is like an A because it has no pockets on right or left; the fifth shows that it is like a V because it has two ends; the sixth shows that it is like an S because it has no junctions; the seventh shows that it is like a D because it has two corners. Combining these models of the letter, determine what it is.

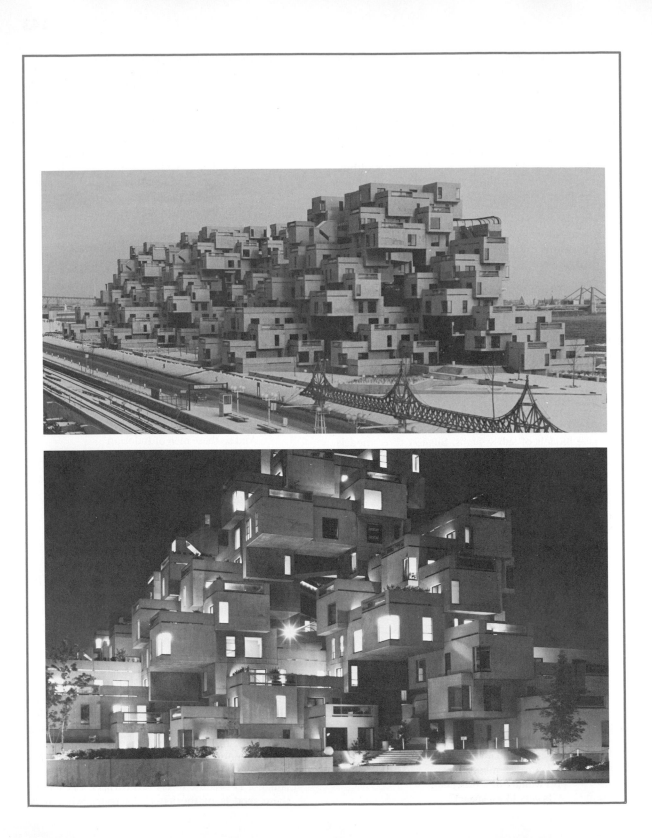

Systems

5

I | INTRODUCTION

One day we read the following news item in the newspapers: "The U.S. Air Force has launched a revolutionary Spy-in-the-Sky which can sweep over military bases in any foreign country, take pictures, and return packages of film on command from ground controllers. It may even operate a television system which could be monitored from the ground!"

On another day we read: "Senator Warren G. Magnuson called today for an international agreement to protect the world's fish supply. He said he will propose a conference for such an agreement when he talks to Russian officials in Moscow later this week."

In a Memorandum of Decision of the Superior Court of the State of California, we read the following: "This is an action for wrongful death. The action arises out of an automobile accident which occurred on May 16, 1960, on a highway known as the Carmel-Pacific Grove Cutoff in Monterey County, California. On the date in question, Don Wells Lyford, a young man sixteen years of age, was driving a 1960 Corvair automobile on this highway and proceeding in the direction of Carmel. The highway involved was a two-lane highway through a wooded section with a number of curves. At the exit of a right hand curve and the beginning of a curve to the left, the Corvair automobile went across the center line and into the opposing lane of traffic and collided with a Plymouth automobile which was proceeding toward Pacific Grove. Don was not thrown from the Corvair but, as a result of the collision, he received certain injuries from which he died before reaching the hospital. Don, the only occupant of the Corvair automobile, made no statement before his death relative to the cause of the Corvair going out of control."*

What do these three stories have in common? They are all news; they are all interesting; they are all unfinished stories; they may have some immediate or long range effect on our lives. But what is important in these stories is that they all involve modeling and they also involve dynamics. We do not attempt to predict the action of a satellite, or the future of the world's fish supply, or the action of a 1960 car, without building a model first.

Habitat, Expo '67, architect
Moshe Safdie. (Photo by CMHC, Ottawa, Canada)

*Memorandum of Decision #771-098, July 29, 1966. The question the Court had to decide was whether the accident resulted from the behavior of the car. As the car switched from the right curve to the left curve, was it impossible to steer properly?

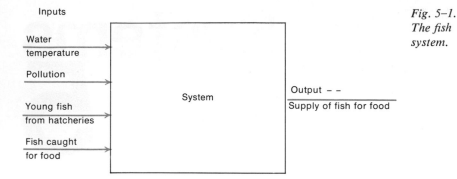

Fig. 5–1.
The fish
system.

Furthermore, all three stories refer to *systems*. A system is a connection of many different parts which interact, often in a complicated way. In a system, several causes combine to give an end result. To understand the system, we have to find the different causes, and find how each influences the total effect.

The world's fish supply, for example, is a system. The number of fish available for food in the future depends on the extent to which we pollute the waters, the number of fish caught each year, the program to grow young fish under protected conditions, and the future changes in water temperature. In other words, we are concerned about the food supply available in the future. We call this end result, or effect, the output signal. There are also many different input signals, or causes, which can influence this output (Fig. 5–1). We can change some of these inputs easily. Others are largely beyond our control.

This example demonstrates why we want to study systems. We try to understand this system—that is, how the output is related to each of the inputs. Once we have such understanding, we can decide how to change inputs in order to give a desired output. What laws should be passed to regulate fishing? What must be done to control pollution? How should we stock the lakes and oceans with young fish?

In this chapter, we will look at various systems. Our immediate goal is to become familiar with a few of the methods which have proven useful in gaining a better understanding of such problems. Our more basic goal is to begin to understand the "systems approach." Recently, economic, social, and political problems have been attacked with systems ideas or from the systems viewpoint.

2 | INPUT-OUTPUT IDEAS

Systems can be discussed in terms of *inputs* and *outputs*. Figure 5–2 shows the input and output for a familiar system. The system is an automobile. The input is the depressing of the gas pedal. The output is the speed of the car.

Certainly we have over-simplified the system if we describe it by the two curves of Fig. 5–2. The speed of the car depends on many signals other than the position of the gas pedal. (For example, if the road is icy, the rear wheels may spin and the speed of the car remain at zero, no matter how the gas pedal is worked.) The two plots of Fig. 5–2 are, however, useful for describing the system under certain conditions. They might describe how a particular car starts up from a stop on a normal concrete road. We might even use such curves to compare different cars (Fig. 5–3).

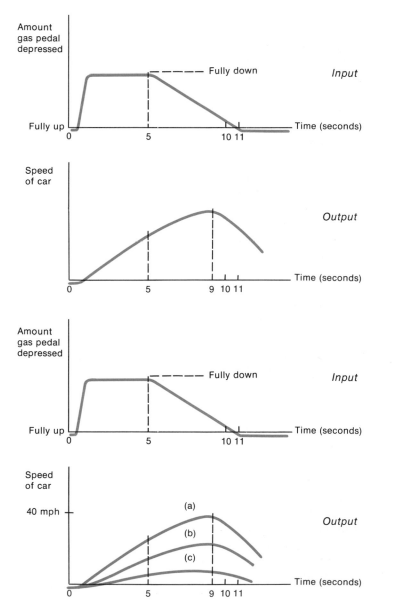

Fig. 5–2.
Signals for
system controlling
automobile speed.

Fig. 5–3.
Comparison of cars
a, b, *and* c. *Car* a *has*
the most "pick-up."

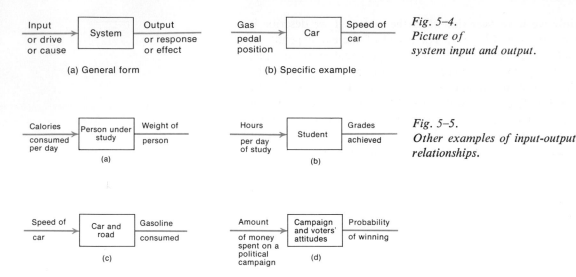

Fig. 5–4.
*Picture of
system input and output.*

Fig. 5–5.
*Other examples of input-output
relationships.*

Thus, in our example, the system is the automobile and the road. The position of the gas pedal is the input signal. This is the signal which moves the system into action. The speed of the car is the output signal, the response of the system to the input, or drive. In other words, when the gas pedal is pressed down, the speed changes; when the system is driven, the system responds.

Rather than write all these words every time we want to describe a new system, we use the simplified notation shown in Fig. 5–4. Here a block represents the system. The input is shown by an arrow pointing into the block. The output is a line coming from the block on the right. This sort of picture is called a *block diagram*. It shows at once the signals to be considered as the input and as the output.

This block diagram is really a picture of the cause-and-effect relationship in a system. The input is the cause, and the output is the effect. In this sense, we can use a block diagram to represent a wide variety of different types of systems. A few of these are shown in Fig. 5–5. We could extend this list to include any cause-and-effect relationship which might exist in nature or among the world of man-made devices.

All the block diagram shows is that there is some relationship between the output and the input. This relationship may not be simple or obvious. For example, in the system of Fig. 5–5d, we sense that the probability of a political candidate winning depends in some way on how much he spends on his campaign (on speeches, billboard signs, television commercials, and the like). It may well be that his chances of winning rise for awhile as his expenditure increases, then fall as the voters become irritated by his over-exposure.

Thus, the block diagram portrays simply what signal we are considering as the input and which signal is the output—the cause and the effect, respectively.

The Block Diagram and the Model

The block diagram defines the system we wish to model. In other words, the block diagram defines the input and output signals. The model must go farther. It must describe exactly how the input determines the output.

The model is *quantitative*. It describes in numbers how the output is determined by the input. Once we have a quantitative model, we can find the output corresponding to a given input. Figure 5–3, for example, is a quantitative model: it shows the way the car speed depends on the position of the gas pedal.

Figure 5–6 illustrates the distinction between a block diagram and a model. The figure is for a very famous example, the model showing the way in which the time required to complete a particular task changes with the number of men assigned to a job. The *complete* figure constitutes the model. Part (a) is the block diagram defining the signals, and (b) shows the *quantitative* relationship between the number of men (N) and the required time (T).

The shape of the N-T curve is familiar to work planners. In building a house, for example, we can reduce the total time required by putting more men on the job. This speeding-up process only works for a while, however. Eventually we reach a point (N_0, T_0 on the plot) after which the addition of more men actually slows down the work.

This slowdown occurs for many reasons. The men may be in each other's way. An excessive effort is wasted in organizing the work and planning each man's activities. The men begin to spend large amounts of time in talking together and enjoying one another's company. Possibly the job is so divided into small parts that the men spend large amounts of time standing around waiting to do their particular parts.

Fig. 5–6.
Model describing effect of changing the manpower assigned to a particular task.

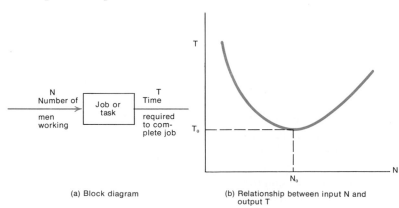

(a) Block diagram

(b) Relationship between input N and output T

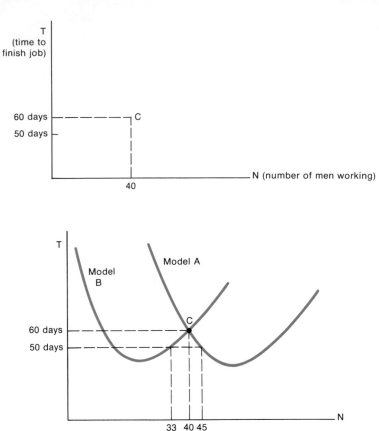

Fig. 5–7.
*Part way through
the job, we estimate
we are at point C. There
are 40 men working. Looking
ahead, we guess it will take 60
days to finish the bridge. Our contract
requires that we be done in 50
days. After that, we have to
pay a heavy penalty for
each day late.*

Fig. 5–8.
*We are operating
at point C (40 men, 60
days). We don't know which
model (A or B) is correct. If Model A
is the one, we should add 5
men. If Model B is correct,
we should lay off 7 men.*

The model of Fig. 5–6 is important in managing a business or a project. It often happens, for instance, that part way through a project we find that the work will not be completed in time. Our bridge may not be completed by the scheduled opening of the entire road. In Fig. 5–7, we estimate the time required will be 60 days, instead of the 50 days which are necessary. As manager of the job what problem do we have? We estimate we will finish in 60 days. We must be done in 50 days. Should we add more men?

Before we can answer this question, we have to know the model. There are two possible models (A and B in Fig. 5–8). If Model A describes our job, we should add 5 more men. If Model B is correct, however, adding more men will require more time. We should reduce the work crew by 7 men.

Unfortunately, what sometimes happens is that more men are added. The project is delayed even more. The manager desperately adds still more men. Finally, the work falls hopelessly far behind schedule.

If the manager understands the model and the two possibilities of Fig. 5–8, he can try adding more men. If it appears that the job is delayed even more, he then can consider reducing the work force. An understanding of the model is important if he is to manage intelligently and successfully. Indeed, he must measure his model and try to find the curve of Fig. 5–6 with the numbers for his particular job. Then he can decide how to finish the job on time and at the least cost.

The block diagram is just a simple way of showing cause and effect. An effect may be the cause for a different system. In other words, the output of one block may be the input to another.

Figure 5–9 shows a simple extension of the earlier example of the gas-pedal car-speed system. Here the total system consists of three parts.

1. The driver. The input is the driver's observation of the time and his location. As these signals change, he decides how fast he must go to reach his destination in time. On this basis, he moves the gas-pedal position (the output for the first block).
2. The car. The input here is the gas-pedal position and the output is the speed.
3. The distance traveled. The speed is the input signal which determines the final output signal: the location, or position, of the car.

Thus, we can put together a number of blocks in a block diagram. The output of one block serves also as the input of the next in a sequence of cause-effect relationships. Indeed, many of the systems which we wish to consider in subsequent chapters consist of a combination of several simple elements, each represented by a block. Figure 5–10 shows another example of a system involving more than one block.

Once we have a block diagram, we can make measurements to find a quantitative model. Thus, the block diagram is just a picture of the different inputs and outputs.

Furthermore, the block diagram shows the train of cause-effect relations. In the thermostat system of Fig. 5–10, the operating

Fig. 5–9.
A more complex block diagram.

Fig. 5–10.
Parts of a home-heating system. When the room temperature drops, the furnace is turned on.

furnace delivers heat. The radiators convert this to a change in room temperature. The input for the total system is the desired temperature (where the thermostat is set). The total system output is the room temperature. In this case, the ideal system would be one which gave an output exactly equal to the input at all times.

If we want to study such a home-heating system, we need to make measurements so we can describe each block in Fig. 5–10 in terms of numbers. How much heat comes from the furnace? What is the resulting temperature change in each room? When does the thermostat go on? How long afterward does the furnace start? It is only after we have such information that we can describe in detail how the system works. The block diagram is just the first step in an understanding of the system.

3 | FINDING THE PARTS OF A SYSTEM

As we shall see at the end of this chapter, the noise environment in which we live is a complex system. There are various inputs, things we want to do which cause noise. For example, trucks carrying food to supermarkets are often noisy. The system output is the noise which surrounds us and which may damage our health.

This noise system is very complicated. The complexity arises because the system has grown without much conscious control by man. Many parts of the system grew without any worry about noise. Only in recent years have we realized that we must control noise, particularly in the cities.

Because of the complexity of the noise system, we will need to look at some simpler systems first. The study of a simple system will suggest how to tackle more difficult problems such as reducing noise in our cities. Now, let's consider the system mentioned earlier, the heating of a home.

What are our goals in this study?

1. We need to understand how the system works in general. How do we heat a house? How do we measure how well it is heated? What causes changes in the living-room temperature?
2. We want to understand what the system is. What are the important inputs and outputs?
3. How are the outputs determined by the inputs?
4. Once (3) is answered, we can ask: How can we improve the performance of the system? Usually, there are two ways to obtain a better output:

   a) We can change the relations between inputs and outputs.

   b) We can control some of the input signals.

Thus, our goal is really to understand the system enough to make decisions about how to improve performance. There are constraints.

Changes may cost money, and we may be limited in the money we are willing to spend. Then our decision problem is to find the best, or optimum, system within the constraints.

With these general ideas as background, let us begin our study of the home-heating system.

How the System Works

The home-heating system is supposed to keep the inside temperature comfortable even when conditions outside vary greatly. In order to have a single example, let's assume that the furnace burns oil. The burning oil heats water to a high temperature (perhaps 190°F). This hot water is then pumped through the house. In each room there is a radiator. The hot water in this radiator heats the surrounding air. The warm air then rises and cooler air moves toward the radiator. The positions of the radiators and thermostat are shown in Fig. 5–11.

One very important part of the system is the thermostat. In our house, there is a single thermostat located on the living-room wall. We are using an inexpensive thermostat which works as follows. The dial is set manually at 72°F. The thermostat includes an element made of two different metals. As the temperature at the thermostat changes, the metals expand different amounts and the element bends. When the room temperature drops below 72°F, the bending is such that a switch is closed and electricity is sent to the furnace and the furnace starts.

Once the furnace starts, the water is heated. As soon as the water temperature rises, the pump starts and heat is sent through the house. In our system, when the temperature reaches 75°F, the

Fig. 5–11.
Floor plan of a
house showing the position of
radiators and thermostat.

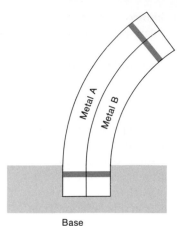

(a) Two different metals are in the shape of strips fastened together at both ends. At a temperature of 70°, both strips are the same length as shown here.

(b) At the higher temperature of 80°, metal A has expanded more than B. Strip A is longer than B. Since the two strips are still together at both ends, the combination must be bent to the right above.

Fig. 5–12.
One possible way a thermostat can operate. The position of the top of the two strips changes with the temperature. Setting the thermostat at 70° means that, as we move from b to a, the switch is closed as we reach position a and electricity is sent to the furnace. There are other possibilities. We can use one metal in the form of a coiled spring. As the temperature rises, the metal lengthens and the spring uncoils.

thermostat turns off the furnace. There is still hot water in the radiators, however, and the room temperature continues to rise until this hot water cools. We find that the heating of the room actually doesn't stop until the room temperature reaches about 77°F.

Thus, on a moderately cold winter day, the living-room temperature changes as shown in Fig. 5–13. With our inexpensive thermostat, the temperature changes over a wide range. At 71° the ladies are putting on their sweaters; at 77° the men are removing their jackets.

Inputs and Outputs of the System

Now that we have a general idea of how the system works, we want to move toward a more refined picture in order to learn what strategies we might use to improve performance. What are the system inputs and outputs? What are the different parts of the system? In other words, what is the block diagram describing the system?

First, what are the *outputs?* What signals measure system performance? Since this is a home-heating system, the output is the temperature inside the house. The system is supposed to keep this temperature constant.

The temperature is not the same throughout the house. The temperature in a bedroom may be very different from that in the kitchen. Which temperatures should we use as system outputs? The answer depends on what we want to know about the system. If we want an overall picture of the house heating, we might choose three temperatures: living room (at the thermostat), kitchen, and one bedroom.

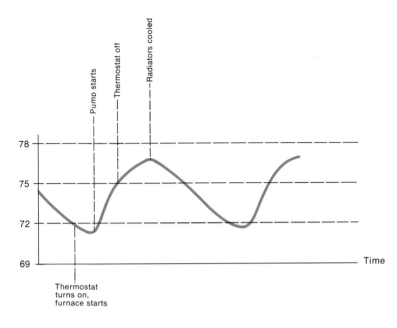

Fig. 5–13.
Temperature at the
thermostat during the day.
(Thermostat is set at 72°.)

Now what are the *inputs?* What causes the three temperatures to change? Here we can list several signals:

1. Outside temperature.
2. Outside wind velocity.
3. Distance the bedroom window is opened.
4. Time the front door is open.
5. Heat from the kitchen appliances (refrigerator, range, and so on).
6. Temperature we set on the thermostat.

At this point, we have to decide what signals to include. There are dozens of different things which affect the inside temperature. We have chosen only six. We have not included the number of people in each room (people give off heat), the windows in other rooms, things causing the air inside the house to move, and additional heat sources such as electric heaters.

In any model, we have to simplify as much as is reasonably possible. We must neglect items which are secondary. After we have finished our study, we can reconsider some of these neglected items. We may find that we have neglected something important. In such a case, we have to start the problem all over again. If we try to include everything, however, the system usually becomes so complicated that there is no hope of ever gaining any understanding.

Once the inputs and outputs are chosen, we can draw a simple block diagram (Fig. 5–14). Figure 5–14 is just a pictorial statement of what we have decided to consider.

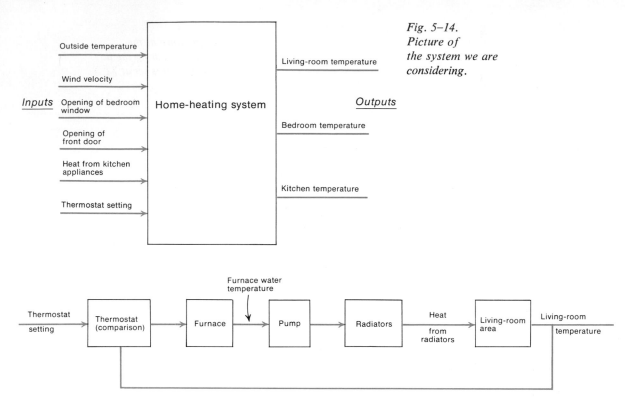

Fig. 5–14.
Picture of
the system we are
considering.

Fig. 5–15.
Block diagram
of the basic heating
system.

Inputs Affect Outputs

Figure 5–14 defines the problem. Now we would like to break this single block down into simpler parts. Our goal is to see how each of the inputs (causes) affects each output (effect).

In order to start, we consider a part of the system. Here we should choose the most important part which is the thermostat setting as input and the living-room temperature as output. The furnace, radiators, and so on are built to allow the thermostat setting to control the living-room temperature.

If we look at just this part of the system, we have the block diagram shown in Fig. 5–15. There are two inputs to the thermostat. We need to compare the setting with the actual living-room temperature. If the comparison indicates the living-room temperature is lower than the thermostat setting, electric power is supplied to the furnace. The pump circulates hot water to the radiators. The heat from the radiators changes the living-room temperature. The temperature change resulting from radiator heat depends on the size of the living room, how much heat leaks out the windows, and so on.

Figure 5–15 is the basic system. Now let's continue to study only the one output (living-room temperature), but let us try to

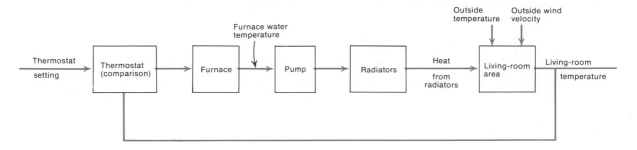

understand what the other inputs do in the system. How do the other inputs (Fig. 5–14) affect the living-room temperature? Can we show these effects by adding to Fig. 5–15?

The outside temperature and wind velocity both affect the system in the same general way. When the outside temperature is very low, large amounts of heat leak out of the living room continually. The same thing happens when a cold wind blows directly on the living-room walls facing the outdoors. Thus, these two inputs represent heat lost from the living-room area. Figure 5–16 shows the new block diagram.

The opening of the bedroom window and the use of kitchen appliances have no effect on living-room temperature. (The kitchen appliances might, but our appliances are fairly far from the thermostat.) The opening of the front door does drop the living-room temperature. Thus, the entire block diagram for determining the living-room temperature is shown in Fig. 5–17.

Finally, we want to show the other two outputs in our block diagram. Each of the other rooms is heated by radiators fed hot water by the same pump system. If we add these parts of the total system, we obtain the block diagram of Fig. 5–18. We now have a complete picture of our home-heating system.

Changing System Performance

Figure 5–18 is a model of the system. Each block shows a cause-effect relation which is a part of the system.

Fig. 5–16.
Block diagram
with two more inputs
added.

Fig. 5–17.
Complete block
diagram for finding living-room
temperature.

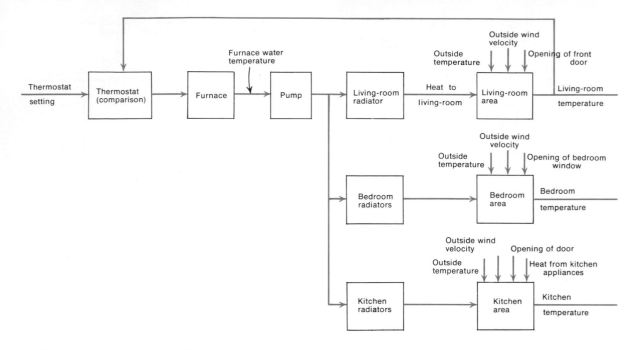

Such a system understanding is valuable if we want to change system performance. For example, perhaps we bought the house last summer and it is now mid-winter. We find that the heating system is inadequate. On cold, windy days, the living-room temperature drops to 60°F regardless of the thermostat setting. What can be done? Figure 5–18 suggests we consider the following alternatives:

Fig. 5–18.
Complete block diagram for six inputs and three outputs.

1. We might try to change the inputs. Four inputs affect living-room temperature: thermostat setting, outside temperature, outside wind velocity, and opening of the front door. We can't do anything about the first two. We might cut the effects of wind by planting dense, high shrubs across the front of the house. We can decide to use only the side door during cold days, so the front door is kept closed.

2. We might try to change the characteristics of individual blocks in Fig. 5–18. Some possible steps are:

   a) Improved radiators. Here we can be sure furniture doesn't block air flow to radiators, and that dust is removed from the radiators. As a last resort, we may decide to add radiators.

   b) Raise the water temperature at the furnace before the pump starts.

   c) Decrease the effects of outside temperature and wind by using storm windows, insulating the walls, and adding heavy drapes across the wall facing the wind.

In the same way, the block diagram indicates steps we might take to change the bedroom or kitchen temperature. For example, if the kitchen is too warm (because of the appliances), we might shut down kitchen radiators. A much more elaborate and expensive solution would be to add a separate thermostat and pump in each of the rooms.

The home-heating system of this section is meant to serve only as an example of the use of block diagrams. When we are trying to understand an unfamiliar system, we first find a block diagram which shows the ways in which the inputs control the outputs. Once this block diagram is drawn, we can see various ways in which we can change system performance.

In a complete study, we would need to take measurements on our system. From these tests, we would find exactly how each input affects each output. We could then predict what would happen to the living-room temperature under specified conditions. We would decide how completely insulated the living room should be to ensure that the system could maintain a desired temperature even with the coldest winter weather.

Thus, the *systems approach* is a way of looking at complex problems or systems. Using the systems approach we try to pick out those features which are important. Then, an understanding of the system suggests what to change to improve the system.

4 | RATE INPUTS

In many cases, we can find the input-output relation in precise terms. For example, we can say the output of a block is four times the input. If we can describe each block in this way, we can find the total system output from any given inputs. Then we can understand in detail how the system operates and how the system can be controlled.

In the examples we will consider, the blocks can be described by three different relations (Fig. 5–19). The first two of these are familiar. The first is called a *scalor*. Here the output is just a constant times the input. In the diagram the constant is called k. This is merely a number (positive or negative). For example, if k is 2, the output at all times is twice the input.

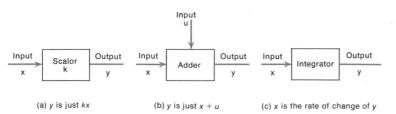

(a) y is just kx (b) y is just $x + u$ (c) x is the rate of change of y

Fig. 5–19.
Three types of blocks for simple systems.

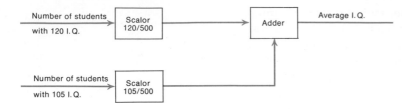

Fig. 5–20.
System for
finding average
I.Q. This block
diagram is just another
way of showing the formula

$$\frac{H \times 120 + L \times 105}{500}$$

where H is the number of
students with an I.Q. of
120, and L is the
number of
students
with an
I.Q. of
105.

The second type of block is an *adder*. The output is the sum of the two inputs.

These two blocks (scalor and adder) are enough to describe some very simple systems. For example, suppose an entering college class is always 500 students. Only two types of students are admitted: those with an I.Q. of approximately 120 and those with an I.Q. of about 105. The average I.Q. of the class is found as in Fig. 5–20. Two scalors and one adder are used.

The third type of block is called an *integrator*. Actually this is just *rate control*. The input is the rate at which the output is changing. Since this block is so important when we model systems, we will consider two examples of the integrator in the rest of this section.

In Chapter 4 we saw that the world population is increasing 2% per year. The population was 3 billion in 1960. Each year thereafter the population was 2% larger than the year before. The resulting population grows exponentially (and doubles about every 35 years).

We can represent this growth in a block diagram as shown in Fig. 5–21. There have to be two inputs to determine the population output. First, we need the population at the beginning. In our case, this is 1960 and the population (p_0) is 3 billion. Then, each year the population grows 2%. The fraction by which it increases each year is $\frac{2}{100}$. This rate is the main input to the integrator. The output is the starting value plus the sum of all the changes since 1960. Thus, integration is just a way of showing that the output depends on the starting value and on the rate at which the output is changing.

Why do we bother to discuss this type of block called an integrator? First, the integrator is important because we can represent many systems by combining integrators with the adders and scalors of Fig. 5–19. Second, the three blocks of Fig. 5–19 are the basic

Fig. 5–21.
Block
diagram for population
growth.

elements of analog computers which are used so often to model and study complex systems.

The integrator takes us from the rate of change of any signal to that signal itself. For example, in a chemical plant where hydrochloric acid is made, the chemical engineer must keep account of the *total* amount of acid in the plant's storage tank. The storage tank is shown in Fig. 5–22. In this figure, the acid flows from the filler pipe into the tank *at a rate of q* gallons per hour. The variation of q with time for a typical morning is shown in Fig. 5–23. (The dip at 10:15 A.M. results from a malfunction of equipment, and the noontime dip is caused by the lunch hour.) At 8:00 A.M. on this day, the volume of acid stored in the tank (the residue from the preceding day) is 1700 gallons. In other words, the initial volume (or initial condition) for the day is V_0 or 1700 gallons.

This system is described by the block diagram of Fig. 5–24. The primary input is q, the rate at which acid is flowing into the tank. The output is the amount of acid in the tank at any time.

In contrast to the population example, the rate q is a signal which changes with time. The amount of acid in the tank at any time is the starting amount plus that which has entered since 8 A.M.

From the curve of Fig. 5–23, we can calculate the total quantity which enters the tank in any period of time. Figure 5–25 shows a one-hour period during which q is constant at 2000 gallons/hour.

Fig. 5–22.
Hydrochloric-acid
storage tank.

Fig. 5–23.
Rate of flow of acid
into tank of Fig. 5–22
on a typical day.

Fig. 5–24.
Block diagram
for hydrochloric-acid
tank.

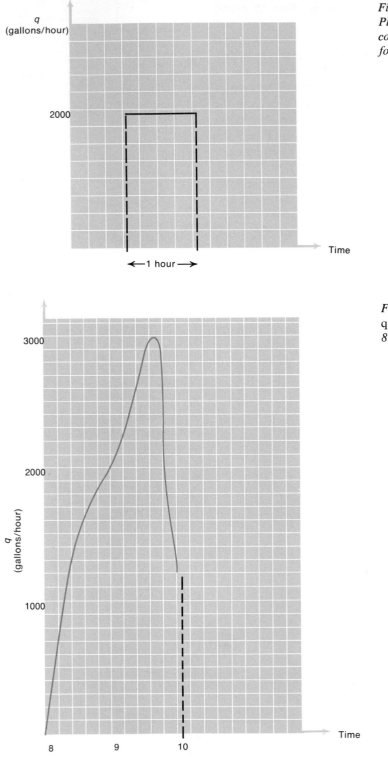

Fig. 5–25.
Plot if q *is*
constant at 2000 gallons/hour
for a one-hour period.

Fig. 5–26.
q *from*
8 to 10 A.M.

During this period, the volume in the tank increases by 2000 gallons. This volume is just the area under the q curve for this period of time.

In Fig. 5–25, this area can be measured by counting the number of squares of the graph paper. There are 32 squares under this plot of q. Hence, 32 squares correspond to a volume change of 2000 gallons (or every 16 squares represents 1000 gallons).

How much acid enters the tank from 8 to 10 A.M.? Now we plot the curve of q versus time (Fig. 5–26). We want the area under this curve from 8 to 10 A.M. If we use the same scales as in Fig. 5–25, every 16 squares corresponds to 1000 gallons. If we count the squares in Fig. 5–26, there are about 120. Hence from 8 to 10 A.M., $\frac{120}{16}$ thousand gallons, or 7500 gallons, enter the tank.

In other words, *the integrator gives an output which is the starting value plus the change. The change is the area under the rate curve from the starting time to the time the output is measured.*

Figure 5–23 shows the *rate* at which acid enters the tank from 8 A.M. on. To find the volume in the tank at 1 P.M., we add 1700 gallons (present at 8 A.M.) to the area under the q curve from 8 A.M. to 1 P.M. (Fig. 5–27).

The preceding example shows that an integrator is essentially an *area-measuring device*. The output at any time (t_1) is the output at the start (t_0) plus the area under the rate curve from t_0 to t_1.

Counting squares on graph paper is a tedious job. Fortunately, there are devices which integrate for us. The simplest is the planimeter (Fig. 5–28). Sometimes the input (rate) signal is an electrical voltage. Then we can use an electronic integrator, a connection of transistors and electrical elements which act as an integrator. We can also integrate with a digital computer. These are all devices which simplify the job of finding areas (or counting squares), and hence permit us to integrate.

Fig. 5–27.
The area of the shaded portion represents the volume of acid added to the tank from 8 A.M. to 1 P.M.

Fig. 5–28.
A planimeter
is a mechanical
device for measuring
the area under a curve.
When the movable
pointer is carried
all around the
boundary of the
region, the counter
reads the area
within the
region. (Keuffel
& Esser
Company)

Initial output
x_0

Velocity
v

Integrator

Displacement or
position x

Fig. 5–29.
Relation
between velocity and
displacement.

Velocity Input

We can easily measure the velocity of an object. This velocity is merely the rate of change of the position. We might describe the velocity of a car as 30 miles/hour on a road due north.*

From the velocity, we can find the position of the car as shown in Fig. 5–29. The present position (usually called displacement) is the initial value plus the area under the velocity curve from the starting time to the present.

Thus, the speedometer and odometer readings of a car are related by a mechanical integrator. The odometer reading is the output of the integrator, and the speedometer is the main input. The change in the odometer reading is the area under the speedometer curve.

In a similar fashion, acceleration is defined as the rate of change of velocity. In other words, velocity is the output of an integrator which has acceleration as the main input. If an object falls freely from a balloon at $t = 0$ with no significant air resistance, the

*
In science, velocity means speed plus the direction of motion.

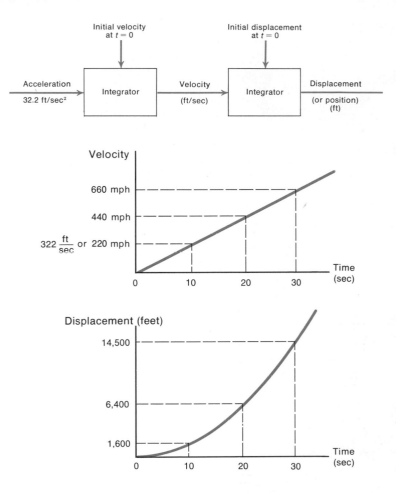

Fig. 5–30.
Two integrators allow
us to find position from a
known acceleration.

Fig. 5–31.
Velocity and
displacement (distance
fallen) when an object falls
from a balloon. We
assume acceleration
is constant.

acceleration is constant, that is, the velocity is changing by 32 ft/sec every second, written as 32 ft/sec² and called the acceleration of gravity. The way the velocity and position of the object change with time are found from the system of Fig. 5–30. The acceleration is constant in our example. The corresponding velocity and displacement are shown in Fig. 5–31.

When a man jumps or falls from a balloon, Fig. 5–31 is not correct. His velocity does not continue to increase forever. If it did, by 34 seconds he would be passing through the speed of sound. We could expect a sonic boom.

What happens to cause the velocity to level off (Fig. 5–33)? As his velocity increases, there is greater and greater air resistance. There is a force tending to oppose this motion. The actual amount of air resistance depends on the shape of the man (there is less when he is curled up). This air resistance grows until it just offsets the gravity, and then the velocity doesn't change. This is called the "terminal velocity."*

*
The air resistance also causes heating. This is the reason for the extremely high temperature of nose cones re-entering the atmosphere from space. In that case, the velocity is already great before the cone enters the air.

Fig. 5–32.
*Free fall in
skydiving. Like a
flock of big dark birds,
skydivers coast earthward
in the French Frog position.
This position is probably
the most popular among
skydiving experts. (Photo
by Bob Cowan,
St. Louis, Mo.)*

Fig. 5–33.
*Actual velocity
of a man in free fall
from a balloon.*

Thus, acceleration, velocity, and displacement are related signals. Velocity is the rate of change of displacement; acceleration is the rate of change of velocity. Figure 5–30 displays how the integrator block shows the relations among these three signals and allows us to calculate from a measured acceleration.

5 | POPULATION MODEL FOR A TOWN

Our three building blocks of the last section were a scalar, adder, and integrator. To illustrate how these blocks are used to represent a real system, we now consider the modeling of the population of a town.

The population of a town changes for four reasons:

1. Births
2. Deaths
3. Immigration (people moving into the town)
4. Emigration (people moving away)

We want our model to show each of these factors.

Births

Usually we specify the births per year by giving the birth rate (the number of live babies born per 1000 population). For example, we might find from town records that the birth rate is 15/1000. If there are 3000 people in the town, we can expect 45 births. In general terms, if we use p for the population and b for the birth rate, the number of new births per year is bp (Fig. 5–34). This bp is the rate at which babies are produced.

Deaths

In a similar way, deaths subtract from the population. The population change is the output of an integrator (Fig. 5–35). The input is the death rate d times the population. This dp is the rate at which deaths occur.

If there were no immigration or emigration, the population p would just be the starting value (p_0) plus the new births minus the deaths. In other words, the system could be modeled by the block diagram of Fig. 5–36. Notice that a scalar block of — 1 follows the death-rate integrator so that deaths are subtracted.

Fig. 5–34.
b *is the birth rate*
per 1000 population and p *the*
population in thousands.

Fig. 5–35.
Population
changes from
death.

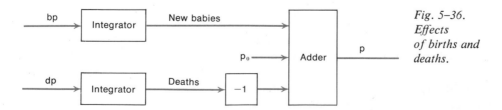

Fig. 5–36.
Effects
of births and
deaths.

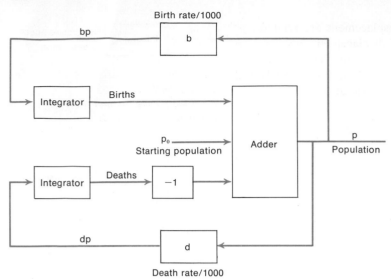

Fig. 5–37.
Complete block diagram
for the town population as it changes
because of births and deaths.

Looking at Fig. 5–36, we notice that the two signals required for the integrator inputs are bp and dp—constants times p. Hence, we can obtain these signals from the output p (Fig. 5–37). We merely add two scalors with gains b and d.

Immigration and Emigration

Finally, we want to include immigration and emigration in our model. If i is the number of people per year moving into the town and e is the number per year leaving, there is a net flow into the town of $(i - e)$. Hence, this $(i - e)$ also represents a rate of change of population. If we put this signal into an integrator, the output is the number of new people each year from migration. Then the total model is shown in Fig. 5–38.

What is the value of a block diagram such as Fig. 5–38? Why is this model any better than a discussion in words of the system? First of all, the model is an exact description of the system. The model shows exactly how the population p is determined from migration, birth, and death rates.

The model is particularly useful also if we wish to study the way p varies with time for several different values of b, d, i, and e. We might program a digital computer to represent the model. We could also use an analog computer described by the same set of equations.

Perhaps, the importance of the model can be best emphasized by considering the problem of a town planning new schools, new police and fire facilities, or new hospitals. Intelligent planning in any of these areas requires prediction of town population several years into the future.

The model indicates that the future population is determined by the present population, p_0, and by the four quantities b, d, i,

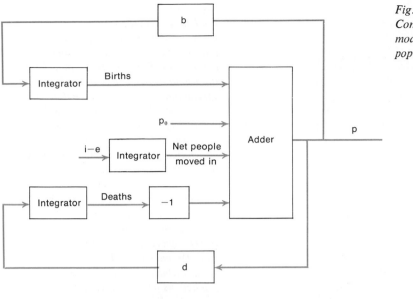

Fig. 5–38.
Complete
model for town
population.

p	population (thousands) at any time
p_0	population at start in thousands
i	immigration rate (thousands of people/year moving in)
e	emigration rate (thousands of people/year moving out)
b	births per 1000 population per year
d	deaths per 1000 population per year

and e. In order to predict p, we must select appropriate values for b, d, i, and e into the future:

b depends on the number of women of child-bearing age and on the economic and sociological characteristics of the community.

d depends on the age distribution of the population, and on the characteristics of public health (epidemics, adequate supply of doctors, accident rate, etc.).

i and e depend on the land available for new housing, the tax policies of the town, the nearby availability of jobs, the quality of the schools, and the general trends in the region (e.g., the extent to which people are moving from a neighboring city to the suburbs).

How do we determine these values? The representation of the model on a computer lets us rapidly find the future changes in p with many different values for b, d, i, and e. Study of these different predictions then allows us to decide on likely future changes in p, as well as the range over which the future population might vary.

Furthermore, the model helps us in setting the tax rate and zoning policies of the town so that in the future there will be enough schools and municipal services for the expanded community. In other words, work with a model of this type allows a much more intelligent approach to city planning and government. The model provides a basis for selection among alternative, possible courses of action.

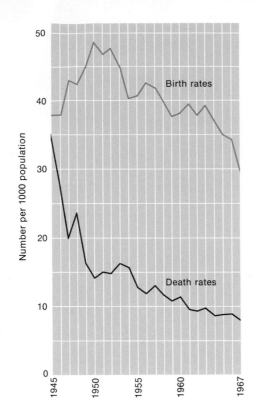

Fig. 5–39
*Finding an accurate model
for a town or country requires
that we insert accurate birth and
death rates. For example, a model for
the island of Mauritius in the Indian
Ocean might be based on the data shown.*

*The population was known to be
700,000 in 1962. If we assume that the
death rate stays constant at 10 per 1000
population per year and the birth rate
levels off at 35 per 1000 population
per year, the net change is +25/1000
population per year (or 2.5% per year).
The population will double in 28 years.*

*A more accurate model would
insert signals for the birth and death
rates (b and d) into Fig. 5–38—
signals which were the best possible
predictions of the future. For
example, the high birth rate
around 1950 should result
in another high value
around 1975, when
girls born in 1950 are
having children.
Thus, to predict
with confidence we
need to know the
population in each
age category.
(Demographic
Yearbook,
United
Nations,
N.Y.,
1954–67.)*

6 | THE NOISE-ENVIRONMENT SYSTEM

For the last fifteen years, the noise in our major cities has been increasing at a constant rate. If this growth continues, every middle-aged resident of New York City could be totally deaf by the year 2000.

If such a statement is true, or even partially "true," we can be certain that New Yorkers will suffer in less obvious ways long before the year 2000. Probably city dwellers are already showing some bad effects of the sound environment. By the time the effects become so obvious that the public demands action, there is likely to be permanent damage which cannot be corrected.

In this section, we want to look at this *noise system*. What are the inputs—the sources of noise? How might these inputs be reduced or controlled? Which are the inputs which are especially harmful to human beings? How can we, as citizens and voters, decide what new laws should be approved to "clean up" the noise environment? How can we understand the problem better?

Where does this example fit into this chapter? We have talked about block diagrams for systems. In both the home-heating system and the town population example, we drew a block diagram. This model showed the cause-effect relations in the system. From the block diagram, we could see which inputs affect which outputs.

Then we could see how to change these inputs to obtain the end result we wanted.

Unfortunately, many interesting systems are very complex. Often there are many inputs and outputs. Even if we know what we would like the outputs to be, we may not be able to decide what to make the inputs.

This complexity often appears in *social systems* (systems which describe problems of society). The system is like a giant balloon. Trying to improve the situation, you change one input. This is like pushing in the balloon at one point. The balloon simply bulges out somewhere else or the pressure inside increases.

Similarly in social systems, an action by city government may improve some part of the system, but actually make things worse in other ways. For example, a city starts a very liberal welfare program to help the people who are unable to work. As a result there is a heavy migration of unemployable people to the city. The generous welfare program becomes much more expensive than planned. Taxes are raised, businesses move out of the city, the number of jobs decreases, and the welfare problem grows steadily worse.

Thus, in social systems, we often cannot draw a neat block diagram. Our system ideas can still be useful. We first study the system. As we gain understanding, we see how to simplify the system. Certain inputs and outputs can be neglected. While quantitative relations between inputs and outputs may be impossible to find, we can try to guess intelligently about which inputs have what effects. From this study, we can begin to understand how to improve the system. The more we understand the system, the more likely we are to be able to make intelligent decisions.

In order to show some of these problems with social systems, we consider the noise environment of a major city. Because it is the largest city, New York happens to be a few years ahead of other cities in most problems. Therefore, some of the following data are for New York, but the problem is very similar even in cities of a few hundred thousand people.

Measuring Noise

If we talk about the noise environment, we must have a way of measuring the amount of noise. (By *noise*, we mean any unpleasant or unwanted sound.) How bad is the noise? How much worse is it now than five years ago? How much worse is it in New York City than in Norwich, Vermont? How bad is it at various times of the day?

These are questions which really cannot be answered easily with numbers. People react differently to noises. This is partly because people have different hearing abilities. But your reaction to noise also depends on your nervousness, your ability to shut out disturbances, and so forth. What we need are some measurable features of noise which indicate its effects on an average person.

Three aspects of noise are important in determining the degree to which the noise disturbs or injures people:

1. *Duration.* The longer the noise lasts, the worse it is.
2. *Loudness.* This is measured in units called *dB* (the abbreviation for decibels, named after Alexander Graham Bell). The dB indicate how loud a sound seems to the human being. A level of 0 dB is the weakest sound an average person can hear. As shown in Table 5–1, normal conversation corresponds to a level of 60 dB. Standing three feet from a riveter results in a noise level of 125 dB.

Conversation	60 dB
Traffic	80 dB
Factory machinery	80 dB
Garbage disposal unit	80 dB
Heavy truck at 25′	90 dB
Food blender	93 dB
Subway train at 20′	95 dB
Power lawnmower	96 dB
Loud outdoor motor	102 dB
Loud motorcycle	110 dB
Hammers, compressors at 10′	110 dB
Large pneumatic riveter at 3′	125 dB

Table 5–1. Decibel levels of common noisemakers.

3. *Frequency.* In Chapter 6, we will see that any signal can be viewed as a sum of sinusoids. The sound signal of noise is no exception. Each noise signal has its own frequencies. Certain frequencies are more annoying than others to human beings. Also, the annoyance depends on the combinations of frequencies. Figure 5–40 shows the frequencies and loudness levels for a few familiar sounds. We see that a boat whistle has frequencies around 200 cycles/second, and the level is about 100 dB. (This is for people standing by the whistle.)

Thus, to describe noise, we have to say how long it lasts, how loud it is, and what frequencies are present.

The effect of noise on a person depends on the duration, loudness, and frequency. The effect depends on the frequency most of all because the human being hears different frequencies differently. Figure 5–41 shows the actual size of a pure note (a sinusoid) at different frequencies when the average person thinks the loudness is the same. In other words, a note at 63 cycles/second must be 55 dB to seem to be as loud as a note of 4000 cycles/sec at 3 dB. (The human ear is very sensitive to frequencies from 2000 to 5000 cycles/sec.)

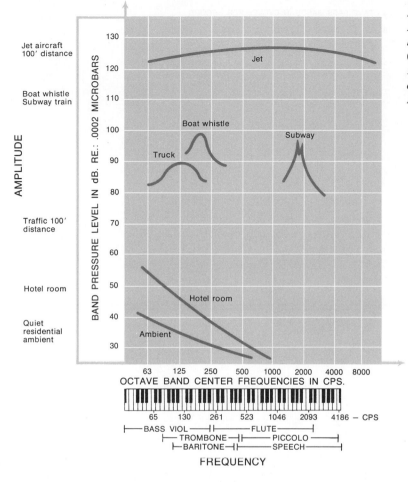

Fig. 5–40.
Noise spectra—
levels vs. frequency.
(*"Toward a Quieter City,"*
Report of the Mayor's Task Force
on Noise Control,
New York, N. Y.)

Thus, a 40 dB noise at 125 cycles/sec would just barely be heard by an average person. A 40 dB noise at 3000 cycles/sec would sound loud. If we worry about the noise environment, our main concern is with noise which has frequencies from perhaps 1000 to 5000 cycles/sec.

To describe the noise environment, we would like a single number. Stating the loudness and frequency of all different parts of the noise is a nuisance. Our single number should measure the noise loudness in each frequency region, and weight this according to how sensitive the ear is for those frequencies. Several measurements of this type are used by engineers. A common one is called dB(A). The A refers to an average or weighted effect. The dB(A) is a single number which describes *very approximately* how loud and annoying a noise seems to a human being.

With this average measurement a jet plane landing 400 feet away has a loudness of 100 dB(A). A truck passing 50 feet away produces noise of 94 dB(A).

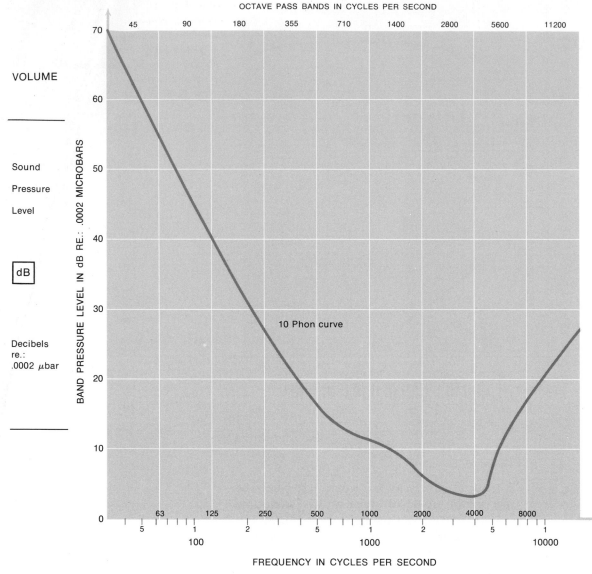

OCTAVE PASS BANDS IN CYCLES PER SECOND

VOLUME

Sound

Pressure

Level

dB

Decibels
re.:
.0002 μbar

BAND PRESSURE LEVEL IN dB RE.: .0002 MICROBARS

10 Phon curve

FREQUENCY IN CYCLES PER SECOND

(FREQUENCY IN Hz [HERTZ OR cps - PITCH])

Effects of Noise on People

How loud can noise be before it disturbs people?

Damage to hearing apparently occurs only when the noise is greater than 85 dB(A). The loudness can, of course, be much greater for short periods of time. Men who work near machinery which frequently causes noise levels above 85 dB should have their hearing checked regularly.

Fig. 5–41.
A curve showing the way the pressure changes as the frequency changes. To a listener, the sound seems to have the same loudness. ("*Toward a Quieter City,*" *Report of the Mayor's Task Force on Noise Control, New York, N. Y.)*

How noise affects people is a difficult question. We can observe that deafness is common when men work in a very noisy environment. (Perhaps 10 million Americans are now seriously affected.) What we are really concerned about, however, is the long-time effect of noise on people. In the United States, hearing ability tends to decrease with age. Studies of African tribes who live in quiet environments have shown that people there still hear as well at age seventy as urban Americans in their twenties.

Other research indicates a need for caution. At the University of Tennessee, the poor hearing of a quarter of the freshmen suggested an experiment. A recording was made of a typical rock music session at a discothèque. For three months, this music was replayed at the same loudness for a guinea pig with one ear plugged—some days for as much as four hours, other days not at all. After 88 hours of music, many of the cells in the unplugged ear were destroyed. There was clear evidence that the hearing ability was seriously damaged by music of normal loudness. Of course, what happens to a guinea pig is not necessarily an indication of the effect on human beings. We need much more research.

Another direction of current research is interesting. Recent studies have shown that noise can move people from a deep sleep to a very light sleep even though they do not awaken. During a night, a sleeping man's brain waves are measured. Every time the man falls into a deep sleep, noise is produced. By morning, he believes he has slept normally, but he actually has not had deep sleep. We find that the man is unusually tired in the morning. Apparently, noise can prevent sleep from being really restful.

Furthermore, noise during sleep may prevent dreams. Major dreams ordinarily last about 50 minutes.* If the man does not have periods this long of reasonable quiet, his dreaming is incomplete. The danger here is not only in the inadequate rest, but also in the fact that people who do not dream might tend to be nervous and perhaps more likely to have emotional problems, possibly because they cannot work out their frustrations in dreams.† Can we foresee universal mental illness as the noise environment worsens?

These comments point out that we just do not understand the effects of the noise environment on people. The few studies that have been made suggest that the effects may be serious. There has been concern that excessive noise may even cause death. We really don't know what noise is excessive or what the effects are.

In such a state of ignorance, we should adopt all reasonable steps to reduce noise. To do this, we need to know the sources of noise (the system inputs). We need to know which sources are most significant in creating the total noise environment. How can people be shielded from this noise? In other words, we need to have a model for the noise-environment system. From this, we can decide where to focus our efforts to reduce noise.

*
The time happens to be exactly the length of the usual class period in college.

†
Most people are unaware they have dreamed during sleep. If you don't remember dreaming very often, there is no reason to be worried.

Fig. 5–42.
The noise
environment.

The Block Diagram of the System

The noise-environment system in a city can be represented by the block diagram of Fig. 5–42. The sources of noise (the inputs) cause various noise levels at different times of the day for our "average" man in the city. These combine to have some effect on his health. We really know almost nothing about the second (right-hand) block in Fig. 5–42. All we can really do is try to control the lefthand block and its inputs.

What are these inputs? That is, what are the major sources of noise for a city dweller? A few of the most important are:

1. Traffic. Here we include the sound of horns as well as of the vehicles.
2. Aircraft. Major sources are helicopters and, especially for those people near airports, jet planes landing and taking off. Sonic booms are occasional problems.

Fig. 5–43.
Some of the major sources
of noise. The chart shows the
levels associated with each source,
compared to the 85 dB considered
safe for hearing. (Photo by Vincent J.
Lopez for Department of Water
Resources. "Toward a Quieter City,"
Mayor's Task Force on Noise Control,
New York City)

3. Construction. During the day, building and road construction equipment is a major source of noise. Compressors and pile-driving equipment are familiar to every city resident.

4. Refuse collection. Garbage trucks are especially noisy. Plastic bags (instead of metal cans) for storing the rubbish on the sidewalk have been a major help.

5. Heating, ventilating, and air-conditioning equipment.

6. Industrial factories and plants.

7. Sirens or alarm systems of emergency vehicles.

8. Subways. Here the tunnels and underground stations trap the noise. The new Montreal system has shown how the noise can be reduced by rubber wheels, better roadbeds, and sound-absorbent materials in the stations. In a very old system such as New York's, the subway is the major place where a large percentage of the population is in a noise environment which is likely to damage hearing directly.

9. Normal social activities. In some dance halls the noise level is 125 dB(A) which is just 15 dB(A) below the level we find standing 30 yards from a jet plane which is landing.

Once we identify these various inputs, we can consider steps to reduce the total noise level at the source (that is, reduce the inputs). Alternatively, if the input can't be controlled, we can try to change the system to reduce the noise levels at the man's ear. For

Fig. 5–44. Some of the noise reductions possible by city-government regulation. Noise from garbage trucks can be reduced 25% if new trucks purchased are designed to be as low-noise as possible. Rubber wheels on subway cars would cut the noise level by 80%. ("Toward a Quieter City," Mayor's Task Force on Noise Control, New York City)

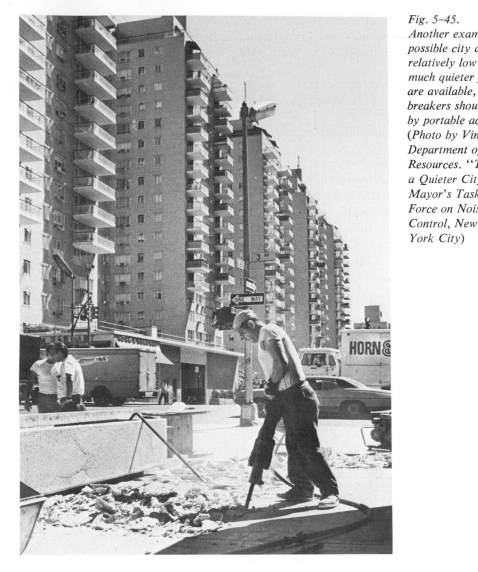

*Fig. 5–45.
Another example of
possible city action (at
relatively low cost). Until
much quieter jack-hammers
are available, pavement
breakers should be surrounded
by portable acoustic panels.
(Photo by Vincent Lopez for
Department of Water
Resources. "Toward
a Quieter City,"
Mayor's Task
Force on Noise
Control, New
York City)*

example, sound insulation in a city apartment can reduce the noise level inside, even if outside noises are uncontrolled. Anyone who regularly rides the New York subways should use ear plugs or ear covers similar to those worn by outside workers at jet airports.

The sound-environment system is a typical example of a social system. It is very complex (many inputs). There are many important parts of the system which no one understands. It is, however, a system which is of great importance to the quality of life in the broadest sense.

Figure 5–42 represents a general block diagram for the system. We could next proceed to describe the system in more detail. We might, for example, look at the noise inputs which affect our city dweller during the night (when he is trying to sleep). How can these be controlled? What would be gained by a window air-conditioner to bring fresh air in (even during winter) without opening the window to bring in street noises?

Unfortunately, in a complex system like the noise environment, we know so little that it is difficult to find a detailed block diagram with numbers in each block. In such a case, we make measurements where possible, and then we have to guess at some of the cause-effect relations. The development of a noise-control program by the city government requires that we move from our general block diagram to a more detailed study, that we consider the costs of possible controls, and finally that we decide on a policy which reduces noise as much as possible within the allowable costs.

7 | MEASURING THE SYSTEM

A system model increases our understanding. From this, in many cases we can decide how to control the inputs or change the system in order to obtain better outputs. Man's control of his environment depends on this understanding. In decision problems, there is always a conflict. We can improve an output by making other outputs worse, by restricting personal freedom in order to change inputs, or by going to the trouble and expense of changing the system. We always want to find the least expensive and least annoying path to the desired output.

The noise-environment example illustrates this conflict in many ways. For example, we can decrease noise by banning sirens on ambulances. The "costs" are more accidents, more injuries to pedestrians, and more time in bringing emergency cases to hospitals. In order to make a logical decision in such a problem, we have to measure the system. If sirens are forbidden, how many more accidents will occur? How much time will be lost in delivering patients to hospitals? What will be the resulting increase in

Fig. 5–46. Another noise-reduction design. Though the aggressive cross-bar tread (D) is not particularly noisy when new, wear creates vacuum cups (E) which produce a high intensity whine persisting after other noise from the truck has subsided.

The tread shown as F has traction characteristics combining advantages of both cross-bar and rib treads. It is relatively quiet throughout its life. ("Toward a Quieter City," Mayor's Task Force on Noise Control, New York City)

deaths and disability? Finally, are these costs less than the value of lower noise level?

Even in this simple example, it is not easy to do experiments or make measurements. Often we have to make the best possible guesses. For example, to learn what a few minutes delay means in bringing accident victims to hospitals, we cannot purposely delay some people selected at random. About the best we can do is ask experienced doctors to estimate the value which should be given to time saved.

Experiments on Social Systems

In finding a model for a social system, it is often impossible to do an experiment to find specific data. In our noise-environment model, how can we determine the average noise from cars on a certain street? We cannot ban cars for a few days. Even if we could, this ban itself would change so many other activities in the city that we would not be measuring normal noise. When we try to make an unusual input to the system in order to measure the resulting outputs, we either change the other inputs or we alter the system. Then we cannot interpret easily the experimental results.

When we are modeling simple engineering systems, experiments are often useful. In the home-heating system of Section 3, for example, we can measure what happens when we don't use the front door at all for a day, when we stop using kitchen appliances, or when we turn off certain radiators. These are clearly defined experiments which allow us to describe the system quantitatively.

Part of the trouble with social systems is that the evaluation of the system depends partly on people's attitudes. Is the system performance annoying or pleasing? If we make a particular change, how much is public satisfaction increased? How much is this worth?

The answer to these questions lies in the subject of psychology. Asking people what they think about a controversial subject is often not adequate. People tend to answer according to what they believe they should say or what they would like to think, rather than the way they actually think. The phrasing of the question is also critical. For example,

> Do you (approve, disapprove) of the President's accomplishments so far?
> Do you (approve, disapprove) of the President's actions so far?

may elicit very different responses. The word "accomplishments" carries the suggestion that desirable things have actually been done.

To test public attitudes, experiments must be designed which try to measure these attitudes without the person realizing he is being measured, but with his responding in a natural way. For example, an attempt was made to measure the relative attitude of city dwellers toward their countrymen and toward foreigners.

This experiment was performed in three different cities (Paris, Athens, and Boston). In Paris three natives and three obvious foreigners were used to make the tests. The tester would pretend to pick up a piece of paper money, stop a passerby, and ask him if he had dropped it. Counter to the impression of some American tourists, Parisians treated foreigners more considerately than their own countrymen. Indeed, Paris was the only city where this was true.

Other, similar tests were also made in this study. Passersby were asked for directions or to mail a letter, store clerks were purposely overpaid to see if they would return the extra money, and taxi drivers were allowed to overcharge and take indirect routes. In mailing a letter, for example, Parisians again were more cooperative than residents of Boston or Athens and also treated foreigners better than their own countrymen.

Other ingenious methods have been devised to measure public attitudes. One is the lost-letter technique. To find the public attitude toward an organization like the Committee for Pure Air, for example, we would write 400 letters:

a. 100 letters to Mr. Williams Barnes, Chairman, Committee for Pure Air, P.O. Box 317, Boston, Mass., 02101.
b. 100 letters to Mr. William Barnes, Chairman, Committee for Better Education, P.O. Box 317, Boston, Mass., 02101.
c. 100 letters to Mr. William Barnes, Chairman, Committee for Overthrow, P.O. Box 317, Boston, Mass., 02101.
d. 100 letters to Mr. William Barnes, P.O. Box 317, Boston, Mass., 02101.

The first group (a) is the test, the (b) group is addressed to an organization which should be considered favorably by most people, (c) is for a group we would assume unpopular, and (d) is just to an individual.

In each envelope, we enclose a letter which seems important. (For example, we might state we will plan to be at the meeting two weeks hence and would enjoy having dinner afterward in order to discuss future actions.) The letters are then sealed and stamped, but not mailed. (Preferably, they are sealed in such a way that we can later tell if they have been opened and resealed.)

The 400 letters are now "lost," one by one, throughout the area where public attitudes are to be measured. Each letter is dropped face up in a spot where it obviously will be seen by people passing by.

A few days later, the mail is collected at Box 317 in Boston. Of the 100 letters in each category, the number returned seems to be a measure of public attitude. The more letters returned, the more the public likes the organization. The more letters that are opened, the more suspicious the public is of the organization.

These interpretations show up in the results of a survey in New Haven, Connecticut, to measure public attitude toward the Communist and Nazi Parties:

Letters to	% returned	Of those returned, % that had been opened
Friends of the Nazi Party	25	32
Friends of the Communist Party	25	40
Medical Research Associates	72	25
Mr. William Barnes (an imaginary name)	71	10

The results indicate what people think about the Nazi and Communist Parties. Apparently they are more suspicious of the Communists.

In this kind of test, we try to measure public attitudes. We hope people will respond according to the way they really think—not the way they feel they are supposed to answer.

8 | FINAL COMMENT

The system viewpoint often points out ways to improve the system. For example, many cities now have poison-control centers. When a person (usually an infant) swallows something poisonous, the mother can call the poison-control center. She describes the substance swallowed. The center operator types this information into a computer, which immediately types out the steps which should be taken (what to give the baby to drink, and so forth). The center operator relays this information to the mother.

Once this equipment is installed, it is easy to make a broader system study. The computer keeps track of all inquiries. The number of poisonings are sent regularly to Washington where national data are collected on each type of poison swallowed. From this study of the total-system inputs, it was found several years ago that many poisonings were occurring when the child found a bottle of baby aspirin. These pills were flavored to taste like candy, and the child would eat the complete contents of the bottle. As a result of this study, a law was passed which sets the maximum number of baby aspirin which can be sold in a bottle. Today you cannot legally buy a bottle containing enough baby aspirin to kill a child.

This system viewpoint is useful in many different areas. In economics, for example, a country's production may be described by an *input-output* model, also called a Leontieff model after the Harvard professor who developed the idea. The needed raw materials, labor, and other inputs are listed vertically at the left. The manufactured products are listed across the top (Fig. 5–47). Each entry shows the input needed for the output of that column.

Cold-war interindustry flows				
	Agriculture	Manufacturing	Household final consumptions	Gross totals
Agriculture		400	200	600
Manufacturing	200		600	800
Household labor and other factors	400	400	___	
Gross totals All numbers in billions	600	800		1,400

Fig. 5–47.
Leontief input-output table. This figure just shows the general form. If we are analyzing a nation's economy, each industry is listed twice, in a row and column. Its row describes allocation of its gross output as inputs for other industries and for final consumption. The column shows inputs needed to produce the output.
(P. A. Samuelson, Economics, *McGraw-Hill, 8th edition)*

An input-output array such as this is the basis for economic planning by the government (it has been used in the U.S.S.R., Israel, and India). Shortages of inputs can be anticipated and production can be predicted. Similar analyses are made by individual manufacturing companies as a basis for economic planning.

The system approach is also used in regional planning. Here the inputs are the future changes in population, jobs and housing, family income, college space needed, hospitals required, transportation needs, and so forth. Each of these systems must be coordinated by a central planning agency which makes sure that housing plans are coordinated with schools, hospitals, transportation, industrial development, etc. The development of one system without coordination with the others would result in chaos.

Thus, in the system approach to complex decision problems, we first find the inputs and outputs. We next find a model—a quantitative picture of the way the system relates these inputs to the outputs. The model of the system is the basis of the decision making.

Questions for Study and Discussion

1. The three news items cited at the beginning of the chapter are examples of dynamic systems. Explain what makes each example a system, and what makes it dynamic.

2. The examples of newspaper accounts of dynamic systems cited on the first page of this chapter were all front-page newspaper stories a few years ago. Look through today's paper for an example of a dynamic system which can be modeled. Explain what makes it a system and what makes it dynamic.

3. What are some steps which a government can take to ward off inflation? How would each of these steps have a negative effect on some segment of the population? List the steps and indicate the segment of the population which would be adversely affected.

4. The following block diagram and its accompanying graph is a model describing the effect of automation on the time needed to do a particular job:

(a) Block diagram   (b) Relationship between input N and output T

Explain the reasons for the shape of the curve.

5. Suppose you are the owner of a small grocery store in the heart of a town. You are losing business to a large supermarket on the outskirts of a business district. You decide that the only chance for survival is to enlarge your store into a supermarket-type establishment.

What are the dynamic characteristics of your business with which you must be concerned in deciding whether to expand or go out of business?

6. Draw the block diagram which represents an in-

put of students attending school and an output of graduates.

7. Could the model of the population of the world be called a system (in the sense used in this chapter)? If so, what are the input and output signals?

8. What is usually meant by the phrase "the Mississippi River system"? Suggest a possible input-output relationship which would make it a system.

9–11 The following diagrams show the input and output of a system, and a graph of output vs. input. From an analysis of the diagrams, tell as much as you can about the systems:

9.

10.

What action on the part of industry probably accounts for the change from (1) to (2)? What probably happened at (3)?

11.

Describe the system and make suggestions for leveling the curve.

Patterns of Change

6

For a dull evening, nothing surpasses being paired with an individual who can talk about only one subject and then without any original or provocative ideas on that topic. The danger of falling asleep while driving is especially great on a modern thruway with little traffic, only gradual curves, and no billboards to arouse our interest or wrath. Every college student is painfully conscious of the effect created by a teacher who lectures (with great authority, perhaps) in a monotone—his voice never raised or lowered, with only an occasional pause for breath. Finally, almost no one is anxious to live in a community devoid of change, a town where no roads or parks are built, where there is no change in inhabitants, and where nothing "exciting" ever happens. In spite of the occasional, whimsical plea of the older generation for a return to "the good old days," very few people really want a life without change.

Change is essential in an interesting, exciting, and challenging life; the promise of change in the future stimulates and drives the individual to personal accomplishment. In the same way, situations of interest in applied science are normally characterized by change. We have to guess when and how changes will occur.

Indeed, we can even go further and state that the really interesting situations are those in which change occurs and in which *the change is unpredictable*. This feature of unpredictability adds enormously to the depth of interest. No one would watch TV if he could predict precisely what pictures would be shown for the entire program duration. Interest is aroused because the future changes are uncertain, unpredictable.*

The idea of *unpredictability* is basic in communication. We only transmit information in the process of communicating if the message is unpredictable. As an extreme example, if the next page of this text included only a thoroughly familiar poem (for example, the words of the first stanza of the Star-Spangled Banner), the first few words would convey some information to the reader—the information that the poem was appearing. Once the reader recognized the poem, the remaining words would be totally predictable and would not convey any additional information.

In studying the man-made world, we are interested in this entire process of communication. Indeed, a major portion of our technological world has been created primarily to enhance the ability of people to communicate among one another. If we are to understand devices to assist in this process of communication, we must understand at least a little of the nature of the information which is conveyed.

In order to measure this information, we start from the idea that the amount of information contained in a message depends on the probability of that message. As an example of how information and probability are related, we consider a message or signal which consists of a sequence of 12 binary numbers, each either a zero or a one. Of all the possible messages we might send, one possibility is

$$0\ 1\ 1\ 0\ 0\ 1\ 0\ 1\ 0\ 0\ 0\ 1$$

Another possible message is

$$1\ 0\ 0\ 1\ 0\ 0\ 0\ 1\ 0\ 1\ 1\ 1$$

and so forth.

When we send this message, we have 12 different decisions to make: for each number in turn we can choose between 0 and 1. The man who receives our message thus obtains 12 different choices between two alternatives.

If for each of these choices, the 0 and 1 are equally probable alternatives (i.e., the receiver has no reason to suspect a 1 was more likely than a 0), we say that each number sent carries *one bit* of information. The bit is the basic unit of information; a simple choice between two equally likely alternatives is one bit. The total message, consisting of 12 successive choices, then carries 12 bits of information.

If we are using an alphabet with more than two symbols (e.g., the English alphabet of 26 letters plus a space, punctuation marks, and numbers), each letter in the message carries an amount of information which depends on the probability of that letter occurring: the smaller the probability, the greater the information. Thus,

*There is an interesting example of the importance of unpredictability. Television stations frequently broadcast sports events (football games, prize fights, etc.) a day or several hours after the actual occurrence. If one has learned the results from the newspapers or news broadcasts, interest is sharply diminished; the more detailed knowledge one has, the less the interest in the delayed telecast.

in a message using the English language, the letter E represents much less information than an X or Z, since E is a common letter (its probability is high).

In this chapter our interest focuses on *changing signals*, signals in which the nature of the change is the primary feature of interest. In particular, we wish to consider a few of the types of change which occur most often in the man-made world, how these changing signals can be used to understand the operation of a system, and finally some of the ways in which systems can be designed to control these changing signals. Our primary purpose in this introductory section is to emphasize that unpredictability and change are essential elements of information. When we communicate, we are interested in sending information to another person or to a machine. The signals we send (by voice, telephone, television, books, and so forth) must involve change.

In writing or speaking, we ordinarily do not want to maximize the information content. For example, if this book contained a maximum amount of information, every letter, word, sentence, figure and so forth should come as a surprise to the reader. Obviously such a book would be totally unreadable; any well-written book leads the reader gently from one thought into the next. This can be achieved only by a certain amount of redundancy—extra words and sentences to lead gradually into a new idea.

The redundancy of the English language can be illustrated very simply as follows. We write a logical paragraph. For each letter (one by one), we flip a coin. If the coin comes up heads, we leave the letter untouched; if tails, we erase the letter. Approximately half the letters are now erased. Now we give the paragraph with erased letters to a friend and ask him to fill in as many of the letters as possible.

The usual success in completing the paragraph suggests that about half the letters are really unnecessary. In a page, even more letters could be omitted, and indeed in a full novel or a textbook, a few pages missing would normally not be very serious. Similarly, when we speak, a large fraction of the sounds are redundant.

2 | COMMUNICATION WITH LANGUAGE

How is information communicated from man to man by the spoken or printed word? What are the properties of these man-made signals? How can we describe such signals? Can we hope to build machines which automatically translate written material into a different language? Can we build machines which recognize written material and give out sounds similar to those which a man would utter if he were reading aloud? Can we build machines which recognize spoken words? Will we ever in the future "dial" the phone merely by saying the number into the mouthpiece?

188

In this chapter, we cannot answer these questions. Many of them are still being argued by experts. We can only look at some of the characteristics of man-made speech signals.

Language

If we want to study some of the properties of human communication, we should first consider the language man uses to express his thoughts. There are between 3000 and 10,000 languages in the world today. The actual number depends on the degree to which different dialects are considered to be different languages.

The five major languages are Chinese, English, Hindi-Urdu (of India and Pakistan), Spanish, and Russian. More than two-thirds of the world's population speak one of these languages. In terms of printed material, English is by far the most important. For example, there are 2400 newspapers in English, compared with only 1000 in Spanish.

Actually, the number of different languages in the world is decreasing. If there are 4000 different languages today, there were more than 6000 different languages in 1800. We can anticipate an even more rapid disappearance of languages used only by small groups as travel and communication improve.

Fig. 6–1.
"Red Man's Wireless,"
a painting by C. M. Russell
at the Heritage Center.
(*National Cowboy*
Hall of Fame)

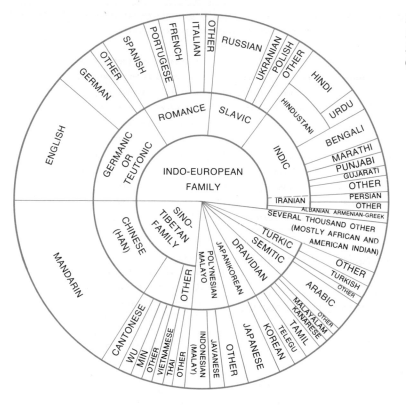

*Fig. 6–2.
Principal
languages of the world.*
(Quinto Lingo)

The ease with which young children learn to speak might suggest that the structure of a language is quite simple. By the age of five, most children speak fluently, although they may have an oversimplified picture of grammar. From this observation we can easily convince ourselves that our language has a logical structure. Perhaps there is a reasonable number of rules which describe the way sentences are ordinarily put together, the way words follow one another, and the way the language is built to match the capabilities of the human brain.

The hope that a reasonably simple structure for language does exist led to the optimistic feeling in 1955 that a few years of research could lead to computers programmed to translate automatically. For instance, we would be able to type an article written in Russian into the computer, the computer would then type out the English version. What information would be required for such automatic translation?

First, we need to store in the computer memory the translation of perhaps 30,000 Russian words.* But a word-for-word translation is certainly not very useful. Many Russian words have several possible English equivalents, and we must have rules for selecting one. Furthermore, word order must be rearranged if the English

*
James Joyce in *Ulysses* uses about 30,000 different words. The average college student has a working vocabulary of 10,000 words, so our figure seems reasonable unless we want to handle many different scientific fields with their specialized vocabularies.

```
P* V NASTO45EE VREM4 MY NICEGO NE MOJEM SKAZAT6 O PRICINAX          IN PRESENT TIME WE NOTHING CAN SAY CONCERNING THE CAUSES
STOL6 SIL6NYX RASXOJDENII MEJDU 3KSPERIMENTAL6NOI I                 OF SO STRONG DERIVATIONS BETWEEN EXPERIMENTAL AND THE
RASCETNYMI VELICINAMI *                                             CALCULATED QUANTITIES .
S TOCKI ZRENI4 SOZDANI4 ISTOCNIKA VYSOKOZAR4DNYX IONOV 3TO          FROM THE POINT OF VIEW OF CREATION OF SOURCE VYSOKOZAR4DNYX
RAZLICIE OKAZYVAETS4 PRI4TNYM , T. K. OBLEGCAET DOSTIJENIE          IONS THIS DIFFERENCE TURNS OUT TO BE PLEASANT , T. K.
NEOBXODIMYX ZAR4DNOSTEI *                                          FACILITATES ACHIEVEMENT NECESSARY ZAR4DNOSTEI .
TAK , NAPRIMER , REAL6NO SOZDAT6 ISTOCNIK S INTENSIVNOST6H          THUS , FOR EXAMPLE , REALLY CREATE SOURCE WITH INTENSITY
2.10,SUPER, 13 IONOV URANA ZAR4DNOSTI PLUS,24 V SEKUNDU ,          2.10,SUPER, 13 THE IONS OF URANIUM ZAR4DNOSTI PLUS,24 IN THE
KOTORYE MOGUT BYT6 USKORENY DO NEOBXODIMYX 3NERGII NA               COURSE OF SECOND , WHICH CAN BE ACCELERATED UP TO NECESSARY
QIKLOTRONE S DIAMETROM POLHSOV 4 M *                                ENERGIES ON THE CYCLOTRON WITH THE DIAMETER OF POLES 4 M .
P* V ZAKLHCENIE AVTORY SCITAHT SVOIM PRI4TNYM DOLGOM                    IN CONCLUSION THE AUTHORS COUNT ITS/THEIR PLEASANT LONG
POBLAGODARIT6 AKADEMIKA G.N.FLEROVA ZA INTERES K RABOTE I           TO THANK ACADEMICIAN G.N.FLEROVA FOR INTEREST TO WORK AND
E.D.VOROB6EVA ZA POSTO4NNUH PODDERJKU I POMO56 , VO MNOGOM          E.D.VOROB6EVA FOR THE CONSTANT SUPPORT AND HELP , IN MUCH
OBESPECIVWIE USPEWNOE VYPOLNENIE DANNOI RABOTY *                    WHICH PROVIDED THE SUCCESSFUL COMPLETION OF THE GIVEN WORK .
```

is to be easily readable. (We won't even require that the translation be quality literature.)

Research over the decade after 1955 made one point clear: the structure of a language is very complex. From this work, it became apparent that the optimism of the early 1950's was premature. Truly useful, automatic translation has to wait until we have a much better understanding of language.

The complexity of languages is emphasized by the sharp differences from one language to another—differences which seem to point out major areas in which this man-made device was developed in very different ways by different civilizations.

Consider the way a noun is made plural. In English, the vast majority of plurals are formed by adding *s* after the singular form. There are, of course, exceptions for which we add *es*, change a final *y* to *ies*, or use a totally irregular form (women, fish). Regardless of these special cases, we can generally state that, when a noun refers to more than one, an *s* is added.

In contrast, the Chinese language has no required form change when a noun is converted from singular to plural. A number of languages have three different forms: one for singular, one for referring to two objects, and a third for referring to more than two.

Somewhat surprisingly, these are the only three classes into which languages fall if we compare them only on the basis of plural nouns. Each language has either one, two, or three forms for a noun.

How plurals are formed is just one example of the structural elements of a language and the way the rules vary from one language to another and, also, within a language from one word to another.

Structure of Language

We can see the complexity of language in another way. In order to obtain a picture of language, we need to understand that a sequence of letters making up a message is at least partially random

Fig. 6–3.
Example of the result of computer translation into English. The Russian typed into the computer is shown on the left. The "English" translation is on the right. Where "its/their" appears, this is merely the dictionary entry and has nothing to do with the algorithm on deciding which should be used. When a Russian word or term appears in the English translation, it means the term was not included in the dictionary and thus is reproduced in Russian rather than English. (Michael Zavechnak, School of Languages and Linguistics, Georgetown University)

CHINESE AND JAPANESE

By Howard Nelson

Chinese and Japanese calligraphy is not as mysterious as it appears to the western eye. The written characters are simply pictures or ideographs which require a certain amount of logic to recognize what they depict.

It is not too difficult to see the trunk and branches in tree 木, three of which make a forest 森. A man 亻 leaning against a tree 休 is resting or taking it easy.

Sun originally was written to show it on the horizon, but as the language developed, it became 日. This plus 本, root or base, forms the compound 日本, which means Japan, Land of the Rising Sun. The logic behind 中 for central or middle is evident. Together with country 國, it becomes China, which once thought it was the center of the world and often called the Middle Kingdom by English travelers.

Similarly 月, the crescent moon evolved into 月, which joined with sun in the same ideograph 明 means brightness or light. With 白 white, it means to understand in Chinese.

人 man is shown walking as a sign of activity while woman 女, whose occupations are more sedentary, is pictured seated. Two women 姦, means gossip and three 姦 lewdness or treachery. Shows what the Chinese and Japanese thought of their women in the old days. Woman and child 好 is the adjective, good.

A man in a cage 囚 is a prisoner and one who lives on a mountain 仙 is a hermit. One can see slopes tapering to the peak in 山. Take a look at the character 車. You can see two wheels, an axle, and a body; in other words, a wagon. The character for rain 雨 shows drops falling from heaven through the clouds.

Mouth, originally and now 口 figures in every character that has to do with speech. To talk is 言, or vapors or steam coming from the mouth. The early eye, and now, 目 is part of ideographs having to do with vision such as 見, to see.

宀 is a roof with chimney and eaves. Put a pig 豕, recognizable by its bristles, under it, and you have 家 house or home. In China, family and livestock live under the same roof. A hundred 百 people 亻 under one roof is a hotel or inn 宿. A woman in the house 安 denotes peace.

A rickshaw, properly jinrikisha, is a man 人 power 力 cart 車. Judo is the easy 柔 way 道, and hara-kiri, called by the Japanese *seppuku*, is cut 切 belly 腹. Other familiar words are kimono, a wear 着 thing 物, and geisha, an art or gifted 藝 actress or entertainer 者. A big 大 wind 風 is a typhoon and a big 大 lord 君 is a tycoon.

There are pitifully few ways to insult anyone in Chinese or Japanese. The worst thing you can call a Chinese is a *wang-pa*, forget 忘 (composed of the elements dead and mind) and eight 八, or one who ignores the eight commandments laid down by Confucius. Horse 馬 and deer 鹿 unite to become *baka*, a word which the Japanese berate each other for being fools or simpletons.

The Chinese and Japanese can read each other's writing, but only partially. Although they use the same ideographs, grammar and word order are completely different. Surprisingly Chinese grammar is the simplest of all civilized languages while Japanese is one of the most complicated.

To Japanese, the United States is the rice country 米國. To the Chinese, it is the beautiful country 美國.

Fig. 6-4.
Oriental calligraphy.
(TWA Ambassador)

(or unpredictable). If you can predict exactly the next sentence that follows this one, my writing the sentence and your reading it are obviously a waste of effort and time. Thus, the existence of randomness and unpredictability is necessary if the message carries any information.

We might first consider the kind of written message which would result with complete randomness, each letter equally probable, and each letter completely independent of what went before. The English alphabet of 26 letters is used, and we add a 27th symbol representing the space between words. If these 27 letters are equally probable and we choose a sequence at random, we obtain a message like*

XFOML RXKHRJFFJUJ ZLPW FWKCYJ FFJEYVKCQSGHYD QPAAMKBZAACIBZLHJQD.

Similar sequences can be generated if we have a deck of 52 cards plus two jokers. We decide ahead of time that the jokers

* This and subsequent examples are taken from *Science, Art, and Communication* by John R. Pierce. © 1968, by John R. Pierce. Used by permission of Clarkson N. Potter, Inc.

correspond to a space, the red aces to A, the red twos to B, and so forth. The cards are then shuffled and one is selected to determine the first letter. The *complete* deck is then reshuffled and the second letter is determined. At each stage, the complete deck is used, so that the original probabilities ($\frac{1}{27}$ for each symbol) are preserved in each selection. Because of the difficulty of shuffling a deck perfectly, a long sequence will eventually tend to show uneven symbol probabilities, but the process is so tedious that few people want to continue for too long.

The message sample is obviously not English or any other language for that matter. In order to obtain something that begins to look like our language, we might attempt to perform the same experiment, but this time we use each letter according to its actual probability of occurrence in English. Table 6-1 shows these letter probabilities. The table simply means that an E occurs $\frac{107}{1000}$ of

Symbol	Probable number of occurrences in 1000 letters	Probable number of occurrences in 1000 letters if space is not considered
Space	187	——
E	107	132
T	85	104
A	66	82
O	65	80
N	58	71
R	55	68
I	51	63
S	50	61
H	43	53
D	31	38
L	27	34
F	24	29
C	22	27
M	20	25
U	20	24
G	16	20
Y	16	20
P	16	20
W	16	19
B	11	14
V	7	9
K	3	4
X	1	1
J	1	1
Q	1	1
Z	1	1

Table 6-1.
These numbers can be found by counting the letters in many different samples of English writing. We then calculate how many times each letter occurs in every thousand letters, on the average. To get a true picture, we would want to use typical samples of books, magazines, newspapers, letters, advertisements, and so on. (*Reprinted by permission of Curtis Brown, Ltd. Copyright © 1942, 1970 by Inga Pratt Clark. The table we have used is an approximation derived from Pratt's table which gives the probabilities to more significant figures. For example, the probability of a Z is 63/100,000. For our purposes, it is easier to round off these numbers.*)

the time (about $\frac{1}{10}$ of the time). Z, at the other extreme, occurs on the average only once in every 1000 letters. If we now write a message based on each of the letters appearing with its correct probability, we obtain a sample like:

OCRO HLI RGWR NMIELWIS EU LL NBNESEBYA
ALHENHTTPA OOBTTVA NAH BRL.

While there are no recognizable words, the letters seem less fantastic than in our earlier example.

We can generate a new sequence of this sort if we have a table of random numbers. (Table 6–2 shows a sample page from a book of random numbers.) Before opening the book to select a page, we decide how we will read the page. We can read to the right, downward, diagonally, any way we choose. Suppose here we decide to read in the normal left-to-right, line-by-line fashion. Furthermore, we will consider the digits in groups of three, the first three, the second three, and so forth. Finally, before selecting a page, we decide how various numbers from 000 to 999 will correspond to particular letters. Since the space occurs 187 times in every 1000, we might choose to let all numbers from 000 through 186

87 35	67 44	51 49	18 98	97 84	75 22	53 29	10 52	26 87	54 92
25 52	29 67	35 99	48 88	40 68	63 68	82 39	38 47	91 39	11 00
87 17	83 31	25 59	87 48	25 80	24 08	81 45	21 32	90 08	44 31
05 04	40 35	72 95	48 56	77 57	63 19	80 16	48 52	06 47	64 98
81 16	09 24	91 71	29 76	54 01	53 47	30 67	62 95	56 58	10 91
54 85	79 88	57 91	11 69	10 22	71 87	24 92	52 64	42 82	78 95
44 78	19 18	35 40	27 66	89 72	21 17	71 69	95 17	97 17	62 60
97 20	98 97	37 33	93 75	18 88	35 85	46 05	07 20	08 17	66 24
98 77	57 51	40 41	76 24	18 54	60 61	79 13	94 57	50 73	89 68
78 12	77 30	83 30	59 28	73 33	47 07	60 07	45 38	82 10	73 19
41 19	70 62	43 46	06 13	22 38	31 18	64 60	07 14	49 16	28 16
70 64	30 55	67 46	95 79	63 66	82 56	67 10	76 77	03 22	42 18
06 56	09 89	68 87	79 19	35 94	66 18	17 94	72 81	72 77	92 39
29 46	18 28	08 88	48 56	49 44	67 82	72 67	28 83	10 26	58 13
42 14	55 51	72 95	29 25	15 18	25 68	48 92	87 16	78 43	17 47
33 75	87 15	15 23	13 79	62 73	76 69	09 77	82 65	72 47	59 56
09 80	99 61	98 08	34 11	88 79	08 32	46 78	33 58	44 16	12 23
98 31	57 50	85 80	53 39	05 92	54 42	29 01	35 23	09 84	96 64
51 70	52 55	83 12	95 02	79 11	49 79	87 95	98 48	88 68	64 77
27 83	61 07	49 05	46 20	35 78	31 34	42 50	68 11	42 14	29 77
78 84	69 15	64 42	92 39	36 08	56 39	35 02	92 78	46 63	82 98
22 12	89 66	49 09	99 10	62 53	19 31	81 83	50 43	37 42	10 00
69 41	59 54	82 72	44 66	64 03	76 59	12 12	41 56	34 90	26 06
54 99	46 54	51 38	59 07	64 21	81 17	88 47	23 05	63 43	08 67
99 91	82 79	92 62	44 24	01 34	45 16	33 56	17 78	42 86	70 94

Table 6–2.
Tables of
random numbers
used to be generated
by looking at census
information and selecting
data which occur at random.
Since 1950, computers have been
used. A very large number (say 27
digits) is squared, another large
number is added, and the last nine
digits of the result are used. The
process is then repeated. ("Tracts
for Computers," edited from the
Dept. of Statistics, University
College London, No. XXIV,
"Tables of Random Sampling
Numbers," (2nd Series) M. G.
Kendall, B. Babington
Smith, Cambridge
University Press,
London, England,
1951, p. 20.)

correspond to a space; the next 107 numbers to an E; the next 85 to a T, and so forth. Then our rules for finding the letter corresponding to each three-digit number are summarized in Table 6–3.

We see that letters can be chosen according to their individual likelihoods, but the resulting message still would not look like English. Perhaps we can do better if we find a message which has both the correct letter probabilities and the correct probabilities of pairs of letters in sequence (called the digram frequencies). A sample message according to these rules is:

ON IE ANTSOUTINYS ARE T INCTORE ST BE S DEAMY ACHIN D ILONASIVE TUCOOWE AT TEASONARE FUSO TIZIN ANDY TOBE SEACE CTISBE.

In other words, this message has been chosen at random, with the use of a table of random numbers, but with correct letter and digram frequencies. We are beginning to get something that occasionally looks like English and is even in part pronounceable. There are several actual English words in this sample message.

The next step would be to form a message which also preserved the appropriate trigram (three-letter) probabilities.

IN NO IST LAT WHEY CRATICT FROURE BIRS CROCID PONDENOME OF DEMONSTURES OF THE REPTAGIN IS REGOACTIONA OF CRE.

In spite of the complexity, we are still far from English.

We can continue this study to understand the complexity of the English language by considering the formation of sets of words according to the appropriate word probabilities, then according to the probabilities of pairs of words, and so on. This sort of exercise leads ultimately to such word collections as,

I forget whether he went on and on. Finally he stipulated that this must stop immediately after this. The last time I saw him when she lived.

where the words are selected with appropriate probabilities for every group of four in a row.

Even though such words are readable, they of course make no sense and convey very little in the way of information or emotion to the reader. They are convincing evidence that the meaning of a sentence depends not only on the grammatical form, but also on the total environment in which that sentence appears.

As a final example of this complexity, Pierce has cited the sentence: Time flies like an arrow. As he points out, these five words may have many different meanings. Three obvious interpretations are:

1. Time passes swiftly by, just as an arrow whizzes by.
2. Time flies (a special breed of flies) are attracted by an arrow.

Number	Letter		
000–186 inclusive	Space		
187–293	E		
294–378	T		
379–444	A		
445–509	O		
510–567	N		
568–622	R		
623–673	I		
674–723	S		
724–766	H		
767–797	D		
798–824	L		
825–848	F		
849–870	C		
871–890	M		
891–910	U		
911–926	G		
927–942	Y		
943–958	P		
959–974	W		
975–985	B		
986–992	V		
993–995	K		
996	X		
997	J		
998	Q		
999	Z		

Table 6–3.

Notice that more than $\frac{1}{2}$ of the letters will be in this group of 5:

More than $\frac{3}{4}$ of the letters are in this group of 10.

More than 90% are in this group of 16.

3. An order: you are to time or clock flies as they go by, just as you would an arrow.

How can the computer, trying to translate the sentence into another language, ever decide which meaning is appropriate? Yet to the human reader, this sentence appearing in the middle of a story would almost certainly be perfectly clear.

3 | SPEECH

The preceding section leads to the conclusion that automatic or computer translation is not likely in the near future. Language is just too complex, but when we turn to speech signals we are a little more fortunate. While the detailed structure of speech is as complex as language, we can measure some very important properties of speech. From a knowledge of these properties, we can hope to make machines which "understand" very simple spoken commands and which recognize the speaker. Already we can speed up or slow down the speech signals so that the talker appears to be speaking faster or slower.

Speaker Recognition

In the summer of 1967, Israel and several Arab nations were involved in an intense, short war. In the midst of the conflict, Israeli officials called together reporters to hear a taped recording claimed to be a radio conversation between President Nasser of the United Arab Republic and King Hussein of Jordan. The speakers were discussing whether to blame the United Kingdom, or the United States, or both for the destruction effected by the Israeli air force. Translation of the alleged conversation reads:

NASSER: Will we say the United States and England—or just the United States?

HUSSEIN: The United States and England . . .

NASSER: By God, I say that I will make an announcement, and you will make an announcement, and we will see to it that the Syrians make an announcement that American and British airplanes are taking part against us from aircraft carriers . . .

HUSSEIN: Good. All right.

Throughout the world, the question immediately arose: Are these really the voices of Nasser and Hussein? In order to determine this, copies of the tape were brought to Lawrence G. Kersta, a New Jersey communications engineer who specialized in voice identification. Kersta compared the particular signals present when the alleged voice of Nasser spoke certain basic sounds and words with a similar analysis of recordings from Nasser's past speeches. Kersta was convinced that the voice alleged to be Nasser was indeed that of the U.A.R. President.

The accuracy of such speaker-identification schemes is a subject of great dispute among scientists, police officers, and computer designers. All agree, however, that if a highly reliable voice identification system were available, it could be of use in the protection of personal privacy such as the discouragement of threatening and indecent phone calls. In the next few paragraphs, we want to see how such a system may become possible.

The Speech Signal

As you listen to a friend talking to you, what signal reaches your ear?

At your ear, when sound is present, there is a rapidly changing air pressure. If we were to place a very sensitive pressure instrument (a microphone) at your ear, we would find the pressure

Fig. 6–5.
Pressure variation during a sound signal.

Fig. 6–6.
Viewing a sound signal.

changing as in Fig. 6–5. The exact way the pressure varies determines the particular sound you hear.

This pressure signal can be measured by the experimental arrangement shown in Fig. 6–6. The microphone is an instrument which changes the pressure signal into an electrical voltage. The voltage between the two wires from the microphone changes exactly as the air pressure changes at the microphone. As the oscilloscope spot moves from left to right on the screen, the voltage moves the spot vertically. Thus, the oscilloscope picture is a graph of the pressure changes with time.

The pressure signal is exceedingly complicated. The pressure rises and falls, but not at all in a regular way. The signal is so confusing, we cannot hope to identify the sounds by looking at a picture like Fig. 6–5. This sort of picture only convinces us that sound is indeed a series of complex, rapid pressure changes.

Fortunately, mathematics comes to our rescue. There is an important concept in mathematics which states that any real signal can be considered as the sum of simple signals. We may need to add up a great many of these simple signals—perhaps so many we never would want to try actually to do the addition, but the important fact is that the breakdown of a complex signal is possible. From this concept, we shall see that we can understand speech sufficiently to allow us to perform such operations as accelerating a speech signal. Furthermore, this concept is basic in our attempts to understand the sound-environment in which people live.

Before we discuss this idea further, we must understand the simple signals which can be added together to give speech, sound, and so forth. These simple signals are *sine signals* (also called *sinusoids*).*

Sine signals

Figure 6–7 is a plot of a sinusoid. The signal increases rapidly at first. The increase slows down until the signal is a maximum at time t_1. It falls until t_3, then starts to rise again. This cycle is

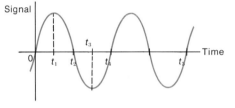

Fig. 6–7.
A sine signal.

*The *Fourier theorem* states that any signal can be considered as the sum of sinusoids (sine signals) of different frequencies and amplitudes. Jean Baptiste Joseph Fourier (1768–1830), orphaned at eight, wrote sermons in his early teens, and became a math teacher at sixteen. In 1798 he traveled to Egypt as an adviser to Napoleon. His experiences there convinced him that the desert heat was healthy. In his later years, he dressed in heavy clothes and kept his room temperature at 90°. It is not surprising that his greatest work was done in the study of heat flow.

repeated. From t_4 to t_5 the variation is exactly the same as from 0 to t_4. Because the signal is continually repeating, we really need to show graphically only one *period* or *cycle* (Fig. 6–8).

There are two important numbers needed to describe our sine signal of Fig. 6–8. We need to state the *amplitude* or how big the signal is. Figure 6–12 shows various sinusoids of different amplitudes. (c) has the largest amplitude. If this were a sound signal, we would say (c) is the loudest of the three signals. In Fig. 6–13, we show a sinusoid of amplitude 8. The second important number which describes the sinusoid indicates the speed with which it passes through a complete cycle or period. To give a number to this idea, we talk about the *frequency* of the signal, or the number of complete cycles per second. In Fig. 6–14 signal (a) has a higher

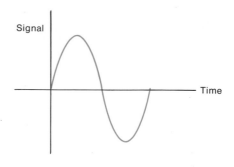

Fig. 6–8.
One period of our sine signal.

Fig. 6–9.
Production of a sine signal. The apparatus shown here can be used to construct a sinusoid. As the wheel is rotated, the pencil moves up and down. If the paper is drawn to the left at a constant speed, the sine signal is traced.

Thus, the sinusoid is the vertical position of P as the wheel rotates at a constant velocity. If the wheel turns at 50 revolutions/sec, the sinusoid has a frequency of 50 cycles/sec. (The Golden Book of Mathematics, Golden Press, N. Y.)

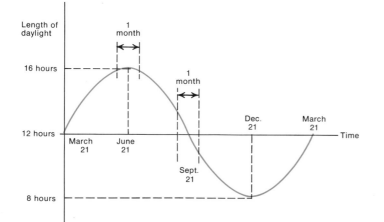

Fig. 6–10.
The sketch shows the daylight hours during the year. (The curve is approximately a sinusoid.) The curve is shown for a location in central United States, where the longest daylight is 16 hours.

The sine signal changes very slowly around the peaks, and rapidly near the zeros. As a result, near June 21 and December 21 the time of sunset and sunrise change very little each day. Around March 21 and September 21, the changes are much greater.

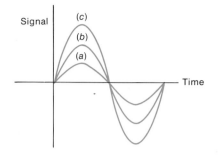

Fig. 6–12.
Sinusoids differing in amplitude or size.

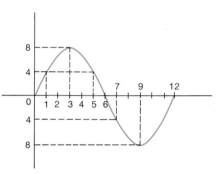

Fig. 6–11. *The tides vary in a sinusoidal fashion at any one location. They are partially caused by the gravitational pull of the moon and sun. They are influenced, however, by the earth's motion and by the motion of the water. There is no theory which allows prediction of how big the tides will be at a particular location or the time when high tide will occur.*

The picture shows Mont St. Michel in France, which is famous for the speed with which the tide changes. High tides rise as high as 50 feet and low tides recede to as far as 11 miles from shore. High tide travels into shore at a rate of 20 miles per hour. The top photo shows high tide. Bottom photo shows low tide. (French Government Tourist Office)

Fig. 6–13. *Proportions of a sinusoid. The peaks (maxima and minima) occur at regular intervals. The zero values occur midway between peaks. During the positive half cycle, for example, the sinusoid has ½ its peak value at moments one third of the time from the zero to the peak in either direction. This particular sinusoid is drawn for an amplitude of 8 and for a period (cycle duration) of 12.*

frequency than either (b) or (c). The frequency of the signal of Fig. 6–15 is 10 cycles/sec. This means that in each second, there are ten complete cycles.

For our purposes, then, a sinusoid can be described by its *amplitude* and *frequency*.

A sine signal in sound is what we call a *pure note* in music. The amplitude measures the loudness, the frequency indicates the pitch.

The terms *frequency* and *pitch* are closely related. The frequency is a term with a mathematical definition. A frequency of 440

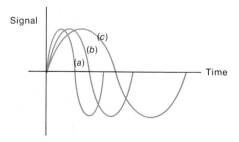

Fig. 6–14. *Sinusoids of differing frequency but the same amplitude.*

cycles/second means that the sinusoid repeats itself 440 times every second. For a sinusoidal sound signal, we can measure the number of cycles precisely and hence determine the frequency.

Pitch is the term used to describe the listener's judgment of the frequency. By international agreement, the musical note middle A is the sinusoid with a frequency of 440 cycles/second. It is not unusual, however, for a symphony orchestra conductor to have his musicians tune their instruments so that middle A is actually a few cycles/second higher. The important thing is that the instruments all be tuned to the same frequency.

Once the frequency for middle A is chosen, all other frequencies are fixed. If middle A is 440 cycles/sec, the A one octave higher is 880 cycles/sec. Each octave up corresponds to a doubling of the frequency (Fig. 6–16).* Four octaves above middle A is 7040 cycles/sec ($440 \times 2 \times 2 \times 2 \times 2$).

Likewise, once the frequency for middle A is chosen, all other notes are fixed (Fig. 6–17). The relationship among the frequencies for the various notes has developed in western civilization as shown in this figure. These are relationships to which our ears are now accustomed by listening to this man-made music from infancy on. When we hear Oriental or Indian music which is based on an entirely different man-made scale, we find it somewhat unusual.

A tuning fork generates a sound wave which is very nearly a pure sinusoid. Vibrating several tuning forks (of different frequencies) does not result, however, in particularly interesting music. The sound is not likely to remind you of any musical instrument. When we play middle A on several different instruments (piano, oboe, violin), why are the sounds quite different? Why don't they all sound like a tuning fork?

There are several reasons. First, and most important, each instrument generates not only the note being played, but also several *harmonics* (also called overtones). Harmonics are signals which are integral multiples of the primary note. For example, if middle A (440 cycles/sec) is being played, the second harmonic is 880 cycles/sec (twice the fundamental frequency), the third harmonic is 1320 (or three times 440), the fourth is 1760, and so forth. Each musical instrument gives a different set of harmonics. If the primary note were middle A with an amplitude of 2 units, two different instruments might give:

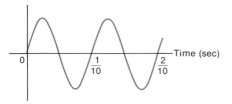

*Fig. 6–15.
Sinusoid with
frequency 10
cycles/sec.*

*
The word *octave* comes from the Latin *octo* for eight. In our musical scale, the eighth note is one octave higher than the first. The octave is divided into 7 parts by the notes.

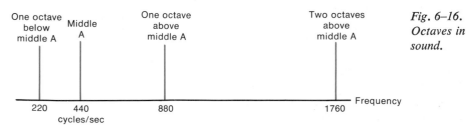

One octave below middle A — Middle A — One octave above middle A — Two octaves above middle A

220   440   880   1760
cycles/sec   Frequency

*Fig. 6–16.
Octaves in
sound.*

Instrument A		Instrument B	
Harmonic	*Amplitude*	*Harmonic*	*Amplitude*
Third	0.5	Third	0.1
Fifth	0.2	Fifth	0.2
Sixth	0.3	Ninth	0.3
Seventh	0.2	Tenth	0.3

The two sounds would be very different to a human listener.

The same note played by different instruments also shows differences in the way the note starts and stops, and the manner in which the amplitude may change during the note. Such characteristics also depend on the person playing the instrument. One of the interesting research programs in recent years has attempted to determine what exactly are the differences between the sounds generated by an expensive violin and by one modestly priced. What features of the instrument cause these differences?

Such research is complicated by the fact that an expert violinist tends to play an expensive, high-quality instrument in a way that is different from the way in which he plays a cheaper instrument. A violin's sound output is particularly sensitive to the style of the player. As the bow is drawn across the string, the string sticks to the bow until it is pulled from its normal position, then slips back, sticks again, and so on. The string vibration is caused by this back-and-forth motion at the point of string-bow contact. The exact way the string vibrates depends critically on the angle between the bow and string, the pressure exerted by the player, the smoothness with which he draws the bow, the rosin used and the way it is put on the bow. Furthermore, to describe in detail any particular note played on a violin, we have to measure many

Fig. 6–17.

Frequencies of the notes which are used in a musical scale. The middle octave of a piano is shown. The standard is A at 440 cycles/sec. A sharp or B flat is the next note up. The frequency is about 1.06 times 440 or 466. B is 1.06 times greater still (or 494).

There are 12 equally spaced notes in each octave. The 13th is one octave above the first.

The white keys on the piano (C, D, E, F, G, A, and B) are the notes used in the C major scale. The black keys are for the five flats or sharps.

Fig. 6–18.

Louis Armstrong blows one of the notes that made him the world's foremost jazz trumpeter.

The curve is a drawing of oscilloscope pictures showing the pressure change when a pure note is played on the trumpet. The signal repeats itself regularly, but is not a sinusoid. The sound is the sinusoid plus a group of harmonics.

The exact signal depends on which harmonics are present and how large the separate harmonics are. These harmonics also depend on who is playing the instrument. (The Bettmann Archive)

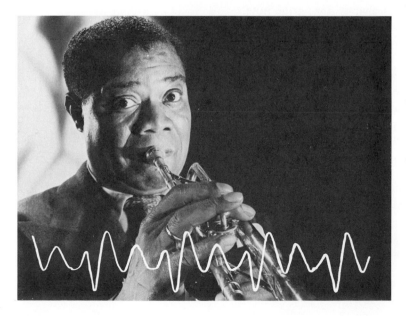

Box 6–1

The existence of harmonics can be demonstrated by a simple experiment. In a piano, each key controls a string which vibrates at the frequency of that note. When the key is idle, the string is clamped so it cannot vibrate. When the key is pressed down, the string is free to vibrate.

If you strike the key for middle C, that string vibrates during the time the key is depressed. The sound which is generated includes several sine signals: one at middle C (about 260 cycles/sec), one at the second harmonic (520), one at the third harmonic (780), and so on. The sound at the second harmonic will excite any string tuned to that frequency and free to vibrate.

To show this second harmonic, you can do the following. Depress the key for the C above middle C (one octave higher or 520) and hold it down. Wait until the sound dies out (eventually the string comes to rest). Now strike *momentarily* the key for middle C. After the middle C note disappears, you will hear the second harmonic note continuing for some time. The 520-cycles/sec string has been excited into vibration by the 260 note.

In a reasonably well-tuned piano and a quiet room, you can detect the third harmonic in the same way (the G one and one-half octaves above middle C).

Fig. 6–19.
An electrocardiogram, measured as the electrical voltage between two points on the body. This is the electrical signal which is associated with each beat of the heart.

The signal shows that there are many harmonics present. (The signal is certainly very different from a sinusoid.)

In recent years, several different techniques have been developed to use a computer to analyze such a signal automatically. From measurements of the various peak values, the durations of the pulses, and the spacings, the computer is programmed to follow the same steps in logical diagnosis as used by a specialist.

different features: the amplitudes of the note itself and each of the harmonics, the way these amplitudes change during the note, and the start and end of the note. More than twenty different characteristics have been used to describe a note. It is not at all clear which of these are important in deciding how pleasant the note is to the human listener. Even if we eventually learn how to describe a single note, we then have the problem of understanding the rapid sequence of notes which occurs when music is played.

The preceding paragraphs begin to point out the complexity of sound. Even in the relatively trivial case of a single note played by a musical instrument, the signal is very difficult to describe. The attempts to program computers to generate sounds which resemble particular musical instruments have really not been too successful. Certainly no one is fooled by a computer which is supposed to sound like a violin.*

Research of this sort is important not only because of the understanding of music, but also because it provides a basis for a fuller understanding of human speech, how man hears, what signals are actually sent from the ear to the brain to allow man to recognize sounds, and how man generates sounds.

*
Recently the French musician Risset generated a "trumpet note" which completely fooled listeners.

4 | SPECTROGRAMS

The most important characteristic of a note played by a musical instrument is the harmonic content: How many harmonics are present? This question is so important, we want a graphical way to display the answer. The display, called a *spectrogram*, is a picture that shows which frequencies are present as time progresses.

The simplest signal is the sound from a tuning fork. Here the sound is very nearly a sinusoid at a single frequency. Figure 6–20 shows the corresponding spectrogram. We plot frequency vertically and time horizontally. At time t_1 the tuning fork starts vibrating at 440 cycles/sec. This sound continues until time t_2, when it is abruptly ended.

In other words, the spectrogram shows the frequencies present at any time. The amplitude, or strength, of a particular sine signal is indicated roughly by how dark or heavy the line is. Thus, Fig. 6–21 shows the spectrogram when middle A is played on a flute; Fig. 6–22, the same note on other instruments.

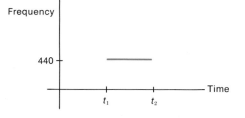

Fig. 6–20.
Spectrogram
of a tuning fork
at middle A.

Fig. 6–21.
Spectrogram
for middle A
on the flute.

Fig. 6–22.
A drawing of a
spectrogram of the sounds of
some common instruments. Depending
on their shape, muscial instruments
emphasize certain harmonics. (From
the book Waves and the Ear *by W.*
A. van Bergeijk, J. R. Pierce, and
E. E. David, Jr. Copyright (c)
1960 by Educational Services,
Inc.; copyright (c) 1958 by
J. R. Pierce and Edward E.
David. Reproduced by
permission of
Doubleday &
Company, Inc.)

Fig. 6–23.
Typical speech
spectrograms. The
spectrograms represent
three different people
saying "man shot Bill." The
first two are males, the third a
female. The vertical scales are
marked in thousands of
cycles/sec. On the
horizontal scale, five
inches represent one
second. (Bell
Telephone
Laboratories)

Spectrograms for Speech

As you speak, you are generating a sequence of air-pressure signals from your mouth to the ears of a listener or to a microphone if your words are being recorded or broadcast. Different sounds correspond to different spectrograms. Each sound you speak has certain frequency components, which change as the sound changes.

Figure 6–23 shows a variety of spectrograms for various speech sounds and different speakers. Obviously, speech represents many very different signals. Can we say anything about the characteristics of speech in general? As a first step, we might look at many different speech spectrograms. If we did this, we would soon notice one important fact. The important frequencies of speech never are greater than 3500 cycles/sec and the fundamental frequencies (not the harmonics) are far below 3500 cycles/sec. A bass singer has fundamental frequencies from 75 to 300 cycles/sec. A soprano voice ranges from 230 to 1400 cycles/sec. Just as with a musical instrument, however, the human voice generates not only the fundamental, but also many harmonics. It is these harmonics which cover the range up to 3500 cycles/sec.

Sharing Communication Equipment

The fact that speech spectrograms show frequencies only to 3500 cycles/sec is one of the things which make modern communication possible. A telephone cable from Chicago to Denver, for example, can transmit signals with frequencies to about 2,000,000 cycles/sec.

Fortunately, with electronic equipment we can shift the frequencies of a signal. For example, we can take a conversation (75 to 3500 cycles/sec) and move this to the range 4075 to 7500 cycles/sec. Each component has its frequency increased by 4000 cycles/sec. The result is, of course, not intelligible, but then we can shift back down in frequency after it has been transmitted, so that we can recover the original signal. This process is shown in Fig. 6–24.

Once this idea of frequency shifting is accepted, we can see how 500 telephone conversations can be carried on one cable from Chicago to Denver at the same time. A system with three conver-

Fig. 6–24. Frequencies can be shifted up or down by any desired amount.

Fig. 6–25. Sharing of communication system. (a) Physical arrangement. (b) Frequencies of composite signal.

sations is shown in Fig. 6–25. We must have complicated equipment at each end to separate the 500 conversations. Even so, this equipment is much less expensive than a separate cable for each conversation.

The concept of frequency shifting (the technical term is modulation) is fundamental to modern communications. Exactly the same frequency shifting is used in radio and television broadcasting. As we tune our radio or change the channel selector in TV, we are changing the frequencies which are to be shifted downward to the original sound range.

Actually, AM radio stations usually broadcast a signal which contains original sound signals up to 5000 cycles/sec (music contains higher frequencies than speech). FM radio signals contain original sound signals up to 15,000 cycles/sec (nearly the maximum frequency the average person can hear). A television picture requires a signal which has frequencies out to 4,500,000 cycles/sec. Thus, each type of signal has a different spectrogram. All these systems, however, operate on the same general principles.

Changing Speech Speed

The spectrogram also suggests how the speed of speech can be changed. Suppose we want to tape record the reading of a novel so that blind people can enjoy the book. The average person can read aloud at about 100 words/minute. With practice he can reach 125 to 150 words/minute, although this is exhausting. A blind person would like to listen at 300 words/minute. Is there any way to speed up the reading?

The first thought would probably be to record the reading, then play back the record or tape at a higher speed. Playing a $33\frac{1}{3}$ rpm record at 45 rpm shows what happens with just a $\frac{1}{3}$ increase in playback speed. The "Donald-Duck effect" makes the speech irritating and unintelligible. In the speeding up, each frequency of the sound has been multiplied by $\frac{4}{3}$ (increased by $\frac{1}{3}$).

The spectrogram suggests a solution. A computer first finds the frequencies present in each short segment of the speech. Each of these frequencies is cut by a selected factor, say $\frac{1}{3}$ so the frequencies are now $\frac{2}{3}$ of the original values. The sum of the new frequencies is now recorded again. The recording is played back at $\frac{3}{2}$ the speed (so the frequencies are multiplied by $\frac{3}{2}$ and, hence, restored to their original values). The new speech signal sounds just like the original, but it is $\frac{3}{2}$ as fast. If the speaker was reading at 120 words/minute, the blind person listens at 180 words/minute. Yet he hears the *normal* tone of voice of the speaker. The result is called *accelerated speech*.

In recent years, accelerated speech has been used not only for reading to the blind, but also for recording lectures in college and in scientific work. Acceleration by a factor of 35% allows the listener to decrease the time required to hear the lecture without any real decrease in understanding. Indeed, some early tests indicate that understanding may actually be improved, perhaps because the listener has to pay closer attention.

Uses of Spectrograms

The speech spectrogram can be used in teaching children born deaf to speak normally. The deaf child is unable to copy the sounds spoken by others. As a result, he talks abnormally even though his lip movement may be nearly normal. In teaching such a child to speak, we find another person with a similar voice pitch and record his speech. As the deaf child speaks, his spectrogram is presented beside that of the normal speaker. By trial and error, the deaf child can try to make his spectrogram look like the one of the normal speaker.

This system has not been widely used so far. (At the Detroit Day School for the Deaf, it has been used successfully.) Simpler aids, displaying voice pitch and intensity, have also been used. All these devices help the deaf by giving them information about their own and others' speech. The devices replace, to some small extent, the missing hearing sense.

Spectrograms are basic research tools as we attempt to build equipment to recognize correctly spoken words or commands (e.g., feeding data into a computer by speaking, or telephone dialing by speaking a sequence of numbers), and to develop speaker-recognition systems.

Finally, spectrograms are being used to understand speech more fully. One of the goals is to build a computer which would "speak" normally. Different sound characteristics could be stored in the computer memory. The system could draw upon these to construct speech. The human quality of speech might be missing, but that would offer no major problem.

The new developments in speech spectrograms will help solve some of the difficult problems found in the synthesis of speech by man-made systems.

Fig. 6–26.
The three tracks of epilepsy (complex sums of sinusoids). A crucial first step in the treatment of epilepsy is identifying which of the three main types of seizure the patient has. Doctors accomplish this with an electroencephalograph machine. They attach small, painless metal electrodes to the patient's scalp. The electrical waves given off by the brain are magnified by the machine and written on graph paper. The most frequent and severe epilepsy is grand mal, "the falling sickness," which involves blackouts and convulsions. Petit mal produces momentary blackouts, with facial twitching and a blank expression. Psychomotor epilepsy dims consciousness and often produces brief amnesia.

5 | SIGNALS RELATED TO SINUSOIDS

The sinusoid is a particularly simple signal to generate and to analyze. The sinusoid is important because many other signals are derived from it. In the following paragraphs, we consider radar and sonar as two important, yet relatively simple, signals which are related to sinusoids.

Radar Ranging

Radar* operates on the following principle. A radio signal of very short duration is transmitted. The sending equipment then

*Radar is a word derived from *radio detecting and ranging*. Initially developed during the 1930's to test radio reflections from the ionosphere (the layer of charged particles over the earth), radar became a vital weapon of World War II when it was refined for detection and location of ships and airplanes.

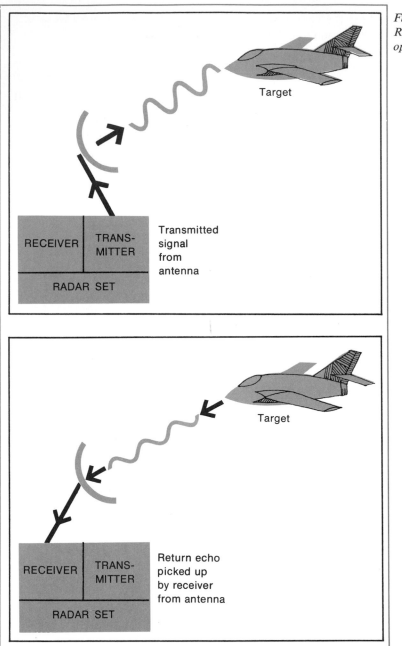

*Fig. 6–27.
Radar system
operation.*

is turned off, and the receiver detects echoes returning from targets which reflect radio waves. Because radio signals travel at constant speed (the velocity of light, which is approximately 186,000 miles per second), the time between transmission of the signal and reception of the echo measures the distance from the radar set to the target (Fig. 6–27).

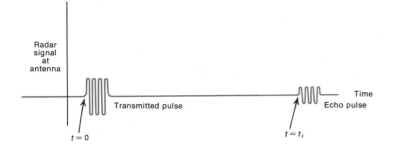

Fig. 6–28.
Antenna signals in radar. The time from the transmitted pulse until the echo pulse measures the distance to the target.

The operation is shown in a different way in Fig. 6–28. At a particular instant of time, which we call $t = 0$, a short-duration sinusoidal signal is transmitted from the antenna. At a later time t_1, an echo signal appears at the antenna. The echo is much smaller than the transmitted signal for two reasons:

1. Of the signal sent out by the transmitter, only a very small fraction hits the target and bounces back. Most of the signal misses the target altogether and either travels through space or bounces back from other targets.
2. Of the signal which bounces back, again only a very small fraction reaches the antenna; most misses the antenna and travels on through space.

As a result, the echo signal typically contains only 10^{-12} times the energy of the transmitted signal.

Even though the echo signal is very weak (just as it is when we listen to a sound echo bouncing back from a distant mountain), the receiver is able to detect this small signal. During the time from $t = 0$ to $t = t_1$, the radio signal travels from the antenna to the target and back again—a total distance equal to twice the range of the target.

The numerical relationship between the time duration t_1 and the range of the target can be derived from the known velocity of light. Because light or radio waves travel at 186,000 miles per second, an echo appearing one second after the transmitted signal ($t_1 = 1$ second) corresponds to a target at a range of 93,000 miles. An echo appearing one microsecond (10^{-6} second or one millionth of a second) after the transmission means the target is at a range of $\dfrac{93,000}{10^6}$ or 0.093 miles or $\dfrac{1}{10.75}$ miles. Thus, every 10.75 microseconds corresponds to one mile of range.

The relationship 10.75 microseconds/mile can be used to convert any measured time t_1 into the corresponding range. For example, an echo delay (t_1) of 90 microseconds means a range of

$$\frac{90 \text{ microsec}}{10.75 \text{ microsec/mile}} = 8.4 \text{ miles}$$

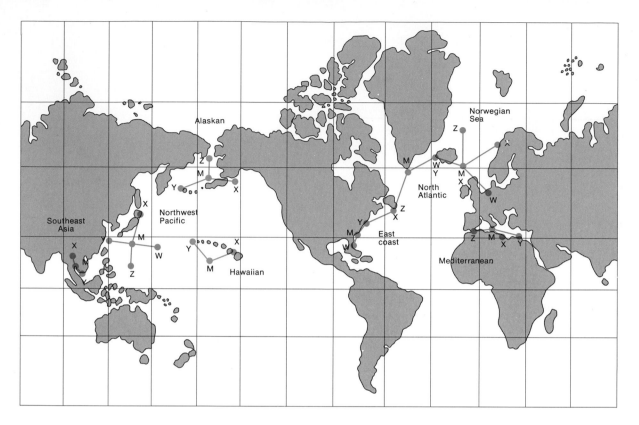

The map shows labeled locations: Alaskan, Norwegian Sea, Northwest Pacific, North Atlantic, Southeast Asia, East coast, Hawaiian, Mediterranean, with stations marked M, W, X, Y, Z.

Radar Direction Finding

The preceding discussion indicates how radar is used to find the range of the target—by measurement of the time that passes before the echo is received. Location of the target requires determination of its *azimuth** and elevation angles. If radar is being used to locate a ship or object on the surface of the earth, it is only necessary to determine the azimuth angle of the target, because the elevation is known to be zero.

The basic technique for bearing measurement involves the use of a highly directional antenna which rotates continuously. A spherical or parabolic antenna "dish" is used with the signal transmitted mostly along the axis of the antenna (Fig. 6–27). As the antenna rotates, short pulse signals are transmitted frequently, so that in any direction several pulses are sent. When an echo is received, we know that the target is in the direction of the antenna at that particular moment.

Radar Measurement of Rate of Change of Range

If the radar system is being used to aim guns shooting at the target, we need to determine the velocity of the target as well as its position. The gun must be aimed to lead the target by the

Fig. 6–29.
Broadcasting stations for the worldwide Loran system. Loran (LOng RAnge Navigation) was developed during World War II to permit ships and airplanes to determine their position. (Aerospace Research, Inc. Updated as of Aug. 17, 1970, by Mr. L. Dennis Shapiro of Aerospace Research, Inc.)

*
The azimuth angle measures the bearing of the target on the surface of the earth clockwise from our reference direction, and the elevation angle measures how much the target is above the earth's surface.

amount the target will move during the flight of the bullet. The situation is identical with the problem of a quarterback passing to an end in a football game. The quarterback throws not to where the end is at the time he throws, but to the point he estimates the end can reach by the moment the ball arrives there.

The target velocity which we must estimate to predict target location when the bullet arrives has two components: the rates of change of the range and of the azimuth angle (again for simplicity we consider tracking a surface target, so elevation is not of interest). Either of these rates of change can be determined by measurement of the signal over a period of time and observation of the way it is changing. In the case of range rate of change, however, there is an alternative approach which is more accurate.

In 1842, Doppler* pointed out that if we move toward or away from a source of light, radio waves, or sound, the observed frequency depends on the relative motion of the observer and source. In particular, if we move toward the source or it moves toward us, the observed frequency of the signal increases. A whistle of an approaching train is higher in pitch than normal.

The reason for this apparent change in frequency is clear if we consider the transmission of sound from the train to our ear. If both the train and we are stationary, the train whistle sends out a sequence of air pressure peaks. Each of these pressure peaks travels toward us at the speed of sound (about 1100 feet/second). Our ear receives or senses the peaks as they pass by. Since we are stationary, the times from transmission to reception for each peak are the same.

If we move toward the train, each successive pressure peak is sensed by our ear a little sooner than it would be if we were further away and stationary. Hence, as we move toward the train, the successive peaks occur slightly closer together. Our motion toward the source tends to make the successive peaks in the sinusoid occur more rapidly; hence the frequency seems to be higher. Furthermore, the amount the frequency is increased measures directly our velocity toward the train. Hence, we can measure the frequency of the radar echo and from this determine the rate at which the target is moving toward or away from us.

Thus, radar can be used for sensing or the determination of the target's range, azimuth angle, elevation angle, and rate of change of range. In addition, the Doppler effect can be used to select (from all the targets spotted by the radar) only those targets which are stationary; more commonly, we are interested in just the moving targets, and we eliminate from the receiver output any stationary targets. For example, in flight control near a city airport, we want the radar screen to show only the moving targets (the aircraft), not the tops of tall, stationary buildings or nearby hillsides.

*
Christian Johann Doppler (1803–1853), an Austrian physicist and mathematician, who was interested in the color (frequency) of light emitted by the stars. The Doppler principle permits astronomers to measure the rotation of stars and planets (one side is moving away from the observer, one side toward the earth, so the colors are different in the light received from the two sides).

Bat Navigation

One of the most impressive sights in the natural world is the exodus at sundown of thousands of bats from Carlsbad Caverns, New Mexico. After spending the daylight hours sleeping in the caves, the bats emerge from the mouth of the caverns, circle a few times, and then head off in search of insects for food. The awesome feature of this mass departure is the uncanny success of the bats in avoiding collisions. To the human observer, the situation seems comparable to that which would occur in the center of a major city if all traffic were able to move at 60 miles/hour with no accidents.

The success of the bat both in navigating among obstacles and in the capture of insects in flight depends upon an extremely refined sonar system.* During normal cruising in search of food, a bat emits short sound pulses about every 0.1 second. During the pulse of transmission, the sound signal is a sinusoid which varies in frequency from about 100,000 cycles/second down to 40,000 cycles/second (Fig. 6–30). The signal sent out is not strictly a sinusoid, because a sinusoid must be of constant frequency when transmitted. Rather it is a frequency-modulated signal (i.e., the frequency is modulated or changed during the transmission). In this case, the frequency is steadily decreased.

Fig. 6–30.
Bat navigation and food location.
a) Signal sent out by a bat.

b) Spectrogram for signal of (a).

Fig. 6–31.
Vikings located enemy shores and fortifications on the fog-shrouded European coast by shouts or hornblasts from their raiding long-ships, judging distance and size by echo return. (Applied Technology, a division of Itek Corporation)

*
Sonar (an acronym derived from *so*und *n*avigation *a*nd *r*anging) was highly developed during World War II for underwater detection (of submarines and of surface ships by submarines). In contrast to radar, sonar uses sound waves, commonly at frequencies slightly above the normal human hearing range (e.g., at 24,000 cycles/second).

Fig. 6–32.
Record made
by a sonar system.
The signal is sent
downward from the surface.
Echoes bounce back whenever
there is a change in sound
transmission. Thus, we can see
fish as the sound signal reflects from
the fish. Also, "fresh" water is different
from salt, so we can see the fresh water
dumping into the harbor from a sewer
outlet. With such records, sonar is used
to obtain accurate maps of a harbor
and also to observe the flow of sewage
lines (many of which were built so
long ago that maps are unavoidable).
(Harold E. Edgarton
and Martin Klein,
M.I.T.)

Presumably, a changing frequency is used so that the bat can recognize the echo from his own signal (particularly when there are thousands of other bats in the vicinity). The frequency change also could permit the bat to recognize precisely which part of the transmitted pulse generated a particular part of the echo.

When the bat discovers a target (an insect), the pulses are transmitted more frequently (as often as 200 pulses per second just before the capture), the frequency of the transmitted sinusoid is lowered (varying from 30,000 to 20,000 cycles/second during a pulse), and the pulses are shortened. Once the target is captured and consumed, the normal cruising operation is resumed.

In the case of the bat, the signals are generated by the vocal cords, and the ears serve as receivers. The location of the target with respect to the straight-ahead axis (i.e., the angle to the right or left) is determined by the comparison of the echo signals received at the right and left ears, just as a human being determines the direction of a source of sound. If the object is in front and to the right of the bat, the echo is received at the right ear slightly earlier than at the left.

The radar signal is a burst or pulse of a sinusoid for a short period of time. The bat's sound signal is a pulse of a "sinusoid" of continuously varying frequency. In both cases, the sine signal is a basic starting point for the understanding of an important class of problems from the natural and man-made worlds.

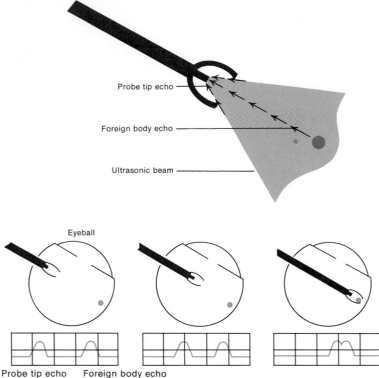

Probe tip echo

Foreign body echo

Ultrasonic beam

Eyeball

Probe tip echo   Foreign body echo

Fig. 6–33.
Surgery by sonar was developed
in 1964 by Dr. Nathaniel Bronson. The
forceps emit a sonar beam which reflects
off the foreign object and the probe tip.
The difference in distance between the
tip of the probe and the object is
indicated on the oscilloscope.
When the two echos coincide,
the surgeon closes the
pincers and
removes the
object.
(Smith Kline
Instruments,
Inc.)

6 | NOT ALL SIGNALS ARE SINUSOIDS

Many signals which occur in the real world are not sinusoidal in character and have very little relation to sinusoids.* While an objective of the chapter is to stress the significance of the sine signal, it is perhaps worthwhile to mention three examples of non-sinusoidal signals.

1. The price of a given stock on the New York Stock Exchange is not a simple sum of a few sine waves. If it were, we could look up the price over a long period of time in the past, evaluate the different sinusoidal components, and use this characterization to predict price variation in the future. While numerous "applied mathematicians" and "engineers" have in the past written articles arguing for such a possibility, it is perhaps sufficient to note that these individuals are still working for a living; none seems to be independently wealthy.

2. Figure 6–34 shows a set of four signals which are very different from sinusoids. These are typical signals measured in research on various safety devices in automobiles. The four signals shown are four different accelerations measured when a stationary car with two passengers is struck from the rear by another car moving at 23 miles/hour.

*
Fourier's theorem, mentioned in Sec. 3, states that any signal can be considered as the sum of sinusoids of different frequencies. If there are thousands of components, such a decomposition is not much help, however.

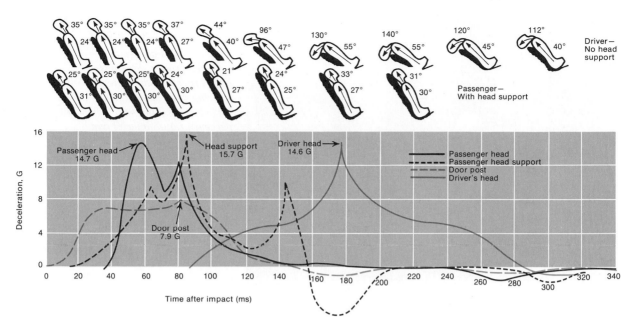

Fig. 6–34.
Kinematics of the
supported and
unsupported head in a
23-mph rear-end collision.
  a) The head of the passenger
    who has a head support.
  b) The head support.
  c) The door post (a part
    of the main frame of the car).
  d) The head of the driver
    who has no head support.
(*Redrawn from "Traffic Safety,*
A National Problem," based on research
by Derwyn Severy, U.C.L.A., The
Eno Foundation for Highway
Traffic Control, Inc.,
Saugatuck,
Connecticut,
1967).

The four accelerations refer to four different measurements within the car which is struck. The dashed line is the acceleration of the door post (i.e., the frame of the struck car). At $t = 0$, the car is hit; shortly thereafter the door post (and car frame) accelerate. The maximum acceleration of about 8g (eight times the acceleration of a body falling freely) occurs about 80 milliseconds ($\frac{80}{1000}$ second) after impact.

The dotted curve is the acceleration of the head support (fastened to the passenger's seat). This part builds up in acceleration a little more slowly than the door post or car body, but reaches the maximum acceleration of almost 16g—about 85 milliseconds after impact. The passenger's head is even slower to start acceleration, but it pretty much follows the head support.

The most interesting curve shown is the curve which represents the acceleration of the driver's head (the −·− curve), since there is *no* head support behind the driver (the test was designed to investigate the value of head supports). We notice that the driver's head does not even start to accelerate until $t = 85$ milliseconds. By that time, the seat (or head support) and hence the driver's torso have already reached maximum acceleration. In other words, the driver's head does not start to move until the driver's torso and the seat have already moved forward a considerable distance (the car was struck from the rear).

The small pictures at the top of the figure show the result. The driver's head tends to remain fixed in space after his torso moves forward. Hence, the head falls backward with respect to the torso.

Fig. 6–35.
A dummy is used
in U.C.L.A.'s accident-
research project. Findings
have inspired collapsible
steering columns and head
supports—and the belief that 70 to 80
percent of auto injuries can
be avoided. (Institute of
Transportation and
Traffic Engineering,
U.C.L.A. Photo
by Bill Bridges)

Fig. 6–36.
Human tracking
experiment (spots move
only horizontally).

When the torso is then slowing down (negative acceleration of the head support), the head is whipped forward. The strong shearing forces associated with the torso pulling out in front of the head cause the severe whiplash injuries.

Thus, the supported head accelerates in synchronism with the shoulders; the unsupported head is accelerated much later by forces transmitted through the neck. When a head rest is not used, severe binding and shear stresses are applied to the spine. The measurement of signals of this sort (on extensively instrumented dummies) indicates the extreme importance of head supports in avoiding serious injury in rear-end collisions.

3. Figure 6–36 shows an experimental arrangement for determination of the dynamic characteristics of a human being in a simple steering or piloting task.

In order to design airplane equipment to allow the pilot to control the aircraft, for example, we need to know how accurately and rapidly a man can steer or pilot. The test arrangement shown is used for this measurement.

The oscilloscope screen shows two spots at all times. One, called spot *R* (for the *r*eference), is moved back and forth horizontally in a random or irregular way. The other spot, called *M* (for the *m*an), is positioned or located as the man directs. The man has a control stick which he holds with his right hand. He can move this stick back and forth across in front of himself. If he moves the stick to the right, the spot *M* immediately moves to the right a corresponding distance.

The man is told to make his spot *M* follow as closely as possible the spot *R* which is moving at random, not under his control. The test arrangement is very similar to the driving-skill game often found at indoor amusement parks.

From experiments such as this (initiated by A. Tustin in England in the 1940's), it was found that the human being can be characterized by:

A time delay (0.2-0.4 second)—the man observes an error, but his response occurs a fraction of a second later.

Proportionality—the man moves his stick an amount proportional to the error between his spot *M* and the spot *R*. In other words, the bigger the error, the more he moves his stick.

Rate measurement—the man measures the rate of change of the error (the rate at which the error is increasing or decreasing) and uses this to predict the future location.

Integration—the man integrates the past error and corrects accordingly.

Thus, the human controller is a complex, dynamic system; his behavior in such a tracking task obeys complex laws or rules which depend on many factors—his degree of fatigue, his motivation to succeed, etc. Once we understand the way the man behaves in such a task, we can design automatic equipment which (for example) assists a pilot to land his airplane at high speed on the flight deck of an aircraft carrier which is pitching, rolling, and yawing in the sea.

In the experimental measurement of human beings in a control task such as just described, the target spot must be moved at random. If the spot is moved sinusoidally, or in a regular pattern of any sort, the man quickly learns the regular nature of the motion and he is able to anticipate accurately the future motion. The experiment then degenerates into a measure of the man's ability to understand the exact character of the motion and to adjust his joy stick accordingly.

7 | PERSON-TO-PERSON COMMUNICATION

In communication with another person, we must use one of our senses to acquire the information. Of the various senses, sight is the most effective for acquiring information. Indeed, perhaps 95% of the total information we receive comes to us through the eyes. Hearing, on the other hand, is the sense which we most often use for direct man-to-man communication through speech (although face-to-face conversation depends strongly on sight as well). Of the other possibilities, the sense of touch also provides the possibility of strong communication between people.

Sight

When we consider the use of sight for communication, the many capabilities and limitations of the eye are important. For example, the eye is relatively sluggish or slow in responding. Hence, movies which show only 24 separate frames, or pictures, each second are seen as a continuous picture, and television pictures which repeat at 60 times/second show no flicker. Actually, our ability to detect flicker depends on the brightness of the picture, but something like 50 pictures per second inevitably looks like a continuous picture. The most annoying flicker rates are those in the range from 10 to 15 cycles/second; possibly in this range there is a relation between the flicker rate and the electrical brain waves.

While the sense of sight is very limited in its ability to detect fast changes, the human being does have remarkable pattern recognition capabilities. A very brief glance at a picture is enough for understanding very complex inter-relationships among the objects. A human being can recognize as many as 2000 different shapes. The reading ability is best when the letters have sharp, pointed corners, and when the letter or symbol height/width ratio is between $\frac{3}{2}$ and $\frac{3}{4}$. This importance of distinct, easily read symbols is familiar to anyone who has looked at different kinds of type when buying a new typewriter.

Our knowledge of the characteristics of the sense of sight has been acquired over centuries of research. During the past two

When in the course of human events it becomes necess
When in the course of human events it becomes nece

When in the course of human events it becomes n
When in the course of human events it becomes n

When in the course of human events it becomes n
When in the course of human events it becomes n

Fig. 6–37. Various kinds of symbols available in commercial publishing. In a book or printed document, particular phrases can be emphasized by the use of distinctive lettering. Comparison of these lines indicates the differences in readability.

*Fig. 6–38.
Eye motion
analysis. The
individual is being
tested to determine how
his eye movement reacts to
a display. On the left is a small
television vidicon camera that shows
the area at which the subject is looking.
On the right is a device that shines a
light on the subject's eyeball and
picks up the reflection that is
superimposed on the field of
view seen by the TV camera.
Experiments of this type
are used in our attempt to
understand the way in
which human vision
collects information.
(Rome Air
Development
Center Display,
Rome, N. Y.)*

decades, research has been particularly intensive as we attempt to understand how different visual pictures stimulate particular nerves, and then how this information is coded for transmission to the brain.* Some indication of the complexity of the problem is given by the following situation:

Printed English contains about 10 bits of information per word.† We can read about 200 words/minute. In other words, in reading we acquire information at the rate of 2000 bits/minute or 33 bits/second. Yet commercial television signals can carry more than four million bits of information per second. Is our television system inefficient? Is most of the information capacity wasted? If we could design a television system well matched to the human eye (that is, a system carrying only information the human eye could use), we could have 4 million/33 or 120,000 times as many TV channels as we now have.

Hearing and Touch

In comparison with the eye, the ear is less accurate in detecting the direction from which a signal comes, but much better in noticing very small changes in sound. The ear can detect a sound break only $\frac{1}{3000}$ of a second long. For detecting brief signal changes, the sense of touch is somewhere between hearing and vision. The human being is able to detect an interruption of only $\frac{1}{100}$ second in a steady pressure applied to the skin.

Thus, we might communicate with a man by attaching to his skin buzzers or pressure devices at perhaps a dozen points around the body (points well separated and where the skin is sensitive to pressure). By applying different signals at the various locations, we can develop a complex code for sending information to him.

*
Much of this present knowledge is described in highly readable terms in the book *Eye and Brain*, by R. L. Gregory, World University Library, McGraw-Hill Book Co., New York, 1966 (available in paperback).

†
A *bit of information* was defined at the very beginning of this chapter. One bit is the information in a choice between two equally likely possibilities (for example, a 0 or a 1).

Such a form of communication is particularly important for people who are both blind and deaf. During the period of 1963 to 1965, there was a major epidemic of German measles in the United States. As a result, several thousand babies were born both deaf and blind. The sense of touch provides the only really useful method of communication with these young people. Communication by touch is also useful in special situations when we are working in darkness and in very noisy environments.

Actually, this idea of communicating through the sense of touch is hardly new. It was suggested by Jean-Jacques Rousseau in 1762. It is only in the last few years, however, that we have begun to understand the types of touch signals which can be understood by a man and the ways in which signals might be coded and presented to the man for easy and reliable understanding. Experimental systems have been built which give rates of communication

Fig. 6–39. *This young woman, blind since birth, is demonstrating how a new electronic system enables her to "see" a telephone with the skin of the back. The TV camera by her right shoulder picks up the image of the telephone on the distant table and converts it into the pattern of dots shown on the TV monitor in the foreground. Then, hundreds of tiny Teflon-tipped cones (not visible here) vibrate against her back allowing her to feel the dot pattern—illustrated in this photograph by fluorescent paint— and thus perceive the image of the phone. (Ralph Crane,* Life *Magazine ©* Time Inc.*)*

appreciably above those of telegraph transmission, but we still are far from matching the rate of communication which can be achieved by speech.

Other Forms of Communication

We have only touched the surface of this fascinating subject. We have, for example, not mentioned the possibility of communication with a sleeping person. We can now buy tape recorder and microphone equipment which can be set up in the bedroom. An alarm-clock system starts the recorder playing during the night after the man is asleep. Tapes are available which are advertised to overcome alcoholism, improve your willpower in dieting, or increase your self-confidence. A few years ago, there was considerable enthusiasm for the hope that a foreign language could be learned in this way.

Serious scientific research is being devoted to the study of how information can be placed directly into the nervous system where it would travel on to the brain. Each ear, for example, senses sounds through about 30,000 nerves (the visual system is much more complex). If we could determine the coding scheme, we might be able to build electronic equipment which would measure the sound signals and then excite the appropriate nerves directly for people who are totally deaf because of damage to the middle ear.

8 | FINAL COMMENT

In this chapter, we have seen that change is fundamental to communication. Changes are represented by *signals*, quantities which vary as time advances. Whether we are considering human communication or the performance of a system, the information is represented by a changing signal.

As an example of a changing signal, we considered speech. This led us to the important particular signal—the sinusoid. The sine signal is described by its amplitude and frequency.

The sine signal is important for three reasons:

1. Many signals are sinusoids, for at least a short time. The electrical voltage at the wall outlet is a sinusoid at 60 cycles/sec in most United States homes. Musical notes are the sums of a few sinusoids. Radar and sonar work with short bursts of sinusoids.
2. Almost any signal can be considered a sum of sinusoids.
3. We will use sine signals in our studies of portions of the man-made world.

Questions for Study and Discussion

1. Describe one situation which you have observed in the past two weeks in which lack of change resulted in boredom and another in which change resulted in interest.

2. Describe three major changes in your life brought about by technology. Be prepared to discuss these in detail.

3. On page 194 there is the statement "Time flies like an arrow" with three obvious interpretations. Try to devise a short statement which has three or more obvious interpretations.

4. Give an example of how predictability of a situation causes a loss of interest, and how unpredictability produces an increase in interest.

5. Give some examples of how technological improvements in packaging have caused problems in the disposal of solid waste.

6. List the sources of solid waste, other than private homes, within one mile of your house. Has waste disposal from these sources been improved or made worse in the past five years.

7. Why do we need to predict such things as production of solid waste or air pollution for from five to ten years in advance?

8. Does the informed citizen need to know what percent of solid waste is plastic, paper, rubber, etc., in order to be able to understand the basis for decisions on waste disposal made by government officials? Explain the basis for your answer.

9. What does the ability of a bat to find its way in the dark have in common with a newly developed method for removing foreign objects from the eye? Explain how this demonstrates that science and technology do work together to help man.

10. Describe three situations in which sinusoids are used in man-made systems.

11. Report on the latest developments in voice identification as described in popular and/or scientific periodicals.

Feedback
7

1 | A FEEDBACK SYSTEM

A man is standing in front of a table. A pencil is rolling across the table toward the edge. The man reaches out with his hand and grabs the pencil.

The *system* for picking up the pencil has operated successfully!

We probably will not read about the success of the system in tomorrow's newspaper. Launching of a spaceship to Venus is likely to attract more attention. Yet the system for grabbing the pencil is similar to the Venus shot. Both problems are very complex, and we are able to succeed in both because we use *feedback*.

What is feedback? To answer this question, let's construct a block diagram for our man-pencil system. Just as in Chapter 5, we will first decide what are the input and the output. Then we will look for the cause-effect relationships which move us from input to output.

The input is the pencil position. The man has no control over this; the pencil moves across the table. The man is in pursuit. The output is the position of his hand. Our system succeeds after this output has been made equal to the input: his hand is on the pencil. In Fig. 7–2, we have the start of a system model.

Now we have to describe how the system works. The man uses his eyes to measure both the input and the output. He watches the rolling pencil and the location of his hand. In particular, he sees the difference between these two signals. The difference is the error between where his hand is and where the pencil is.

In the block diagram, we show this measurement of error by a *comparator* block. The man *compares* the pencil and hand positions. The result of the comparison is the error (Fig. 7–3). The eyes measure this error and the information travels electrically to the brain. There the man computes what he should do. How should he move his hand? Once a decision is reached, electric signals are sent out from the brain to the muscles of the arm and hand. In other words, the nervous system and brain accept information from the senses (the eyes here), and then decide what action to take to improve the output (Fig. 7–4).

These electrical orders from the brain cause the muscles to contract in just the right way to make the hand move toward the pencil. When the hand is at the pencil, the orders call for the muscles to contract, which will cause the fingers to close and grab the pencil. Thus, the complete system is shown in Fig. 7–5.

We said earlier that this system has *feedback*. From Fig. 7–5, we can now see what we meant by this term.

The output of our system is determined by the measured error. This error is sent to the brain where orders are chosen to change the output. In other words, the output results from the measured error (Fig. 7–6).

But in our system *the error depends on the output*. This is *feedback*. We have really a loop of cause-effect relations:

The error changes the output.
The changing output changes the error.
The error changes the output. And so on.

Fig. 7–1.
Feedback system
used for picking up
a rolling pencil.

Input
Pencil position → Output Hand position →

Fig. 7–2.
Input and
output signals for
our system.

Fig. 7–3.
The system measures the error.

Fig. 7–4.
Block added to show decision making
on basis of error observed.

Fig. 7–5.
Model for man-pencil
system.

Fig. 7–6.
An error causes the
output to change.

Fig. 7–7.
A typical Eskimo man. The
adaptation of man to his environment
is an example of feedback. The adaptation
occurs during growth of the individual.
Adaptation also occurs as the gene pool
of the population is changed over
generations.

The average nose shape is related
to the need to moisten the air that we
breathe in. The Eskimo has a very
narrow nose, compared to the length.

The Eskimo typically is small and
squat, with unusually short limbs. In
order to minimize the loss of body
heat, the Eskimo has a relatively
small surface area. In contrast,
certain African inhabitants are
often tall with very long limbs.

There is a relation in
mathematics which states that
in bodies of the same shape,
the taller one has relatively
less surface area. From this,
we would expect people in
colder climates to be
taller than in warmer. This
relation only holds when
the shapes of people are
the same. (The Bettmann
Archive)

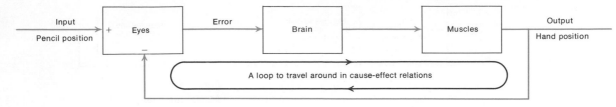

| Input Pencil position | → | + Eyes − | → Error | Brain | → | Muscles | → Output Hand position |

A loop to travel around in cause-effect relations

In other words, a system has feedback when the output depends not only on the input, but also on the output itself. In the block diagram of a feedback system, there is at least one loop we can travel around, always following the arrow direction (Fig. 7–8).

Fig. 7–8.
The feedback in the man-pencil system is shown by the loop.

Feedback Is Important

This chapter has two main purposes. First, we want to understand what feedback is and when it is present. Second, we want to study a few of the remarkable things we can achieve by using feedback.

Even before this study, however, we can see some of the advantages of feedback from our man-pencil system. Why can a man usually grab a rolling pencil? Because the output depends on the error, he is continually measuring the error and making corrections. If his hand starts off too much to the right, he observes this and changes the position of his hand. As his hand nears the pencil, the corrections become more and more refined.

The importance of feedback is clear if we try to do the same job without feedback. (We might blindfold the man.) As the pencil rolls across the table, he moves his hand toward the table, but not necessarily closer to the pencil. System success relies on the continuing measurement of error.

The feedback system automatically seeks the goal of reducing the error to zero: making the output equal to the input. In our man-pencil system, the *goal* is bringing the man's hand to the pencil. In the Venus space shot, the *goal* is to bring the space vehicle to the planet Venus. In steering a car down a highway, the *goal* is to hold the car in the middle of the lane.

In each of these cases the system has a clear goal. There is a measurement of the error: how far is the output from its goal? This error then changes the output toward the goal. This is feedback.

2 | GOAL-SEEKING

Have you ever thought about how many of your actions are directed toward achieving some goal? If you have, you quickly realize that almost everything you do is *goal-directed*, in some sense. These goals may be very simple and immediate. When you insert a dime into a soft-drink dispenser, your goal is to quench your thirst. Goals may be very complex and of long-term significance,

Fig. 7–9.
Mechanization
and automation
are fundamentally
different. Mechanization
uses machines to
replace man's muscles.
Automation replaces the
decision-making activities
of man. In an automated
system, we have feedback.
The output is automatically
compared to its goal, as the
error is measured. On the basis
of this error, a decision is made
how to change the output to reduce
the error.

Mechanization was at the core of the
industrial revolution of the 19th century.
Automation is at the center of the
technological revolution of the 20th century.

Even though automation is a modern
development, there are examples from the
past. The figure shows Leonardo da Vinci's
sketch of a partially automated spit for
roasting meat. The spit was turned by two
separate forces. At the upper right, the
weight falls and unwinds the rope to
turn the spit at about constant speed
(mechanization).

At the bottom, a turbine rotates,
driven by the heat of the fire below.
This rotation is added to the first to
fix the total turning speed of the
meat. As the fire gets hotter, the
hot air from the fire rises more
rapidly. The turbine turns more
rapidly. Thus, the rate at which
the meat turns is increased
automatically as the fire
temperature increases (automation).
Da Vinci's notes contain many
novel examples of both
mechanization and automation
—devices which often were
re-invented hundreds of
years later than his
death in 1519.
(Burndy Library, Inc.)

as when you decide to go to college to prepare yourself for a future occupation.

The purposeful activity of "seeking goals" is of course not unique to men alone, but may be found in all living things. Plants turn their leaves toward the sun and extend their roots toward moist and fertile soil. Salmon battle their way up rivers and through rapids to spawn their eggs in the same creek in which they themselves were hatched. Even the single-celled amoeba moves away from a disagreeable substance to seek a more pleasant environment.

In contrast with such purposeful activity as is exhibited by living things, inanimate objects appear to behave quite differently. Their behavior is based solely on *prior* causes. We do not think of inanimate objects as being influenced by any awareness of what may happen to them in the future.

Inanimate objects appear to differ from living things in another important aspect. With living things we recognize movement toward goals, toward maturity and death, toward satisfaction and reward, toward equilibrium, toward a conscious purpose. We do not think of inanimate objects as seeking "goals." For instance, a rock loosened by the morning's rain falls down the mountainside. We should think it a bit odd if someone suggested that the goal of the rock is to seek a lower elevation. If the rock should strike and kill a snake, only an irrational and superstitious person would assert that this was the rock's goal. Yet if you were to throw a rock at the snake and kill it, we would all agree that it was your goal.

Fig. 7–10.
Example of a feedback system. The hand that holds these scissors is a plastic-and-metal machine complete with motor, gears, strain gauge, and a sensor for measuring the velocity at which the finger closes on the thumb. Information from the gauge and the sensor is used in order to control the hand so that it automatically matches the wearer's intentions (which are detectable as changes in the electrical activity of muscle remnants). This degree of automatic control makes the hand the most sophisticated in general clinical use today.

In this case, the goal is the electrical signal in the arm muscles. The feedback system is the artificial hand, the output is the finger-thumb position. (Photograph Geoffrey Drury courtesy Ministry of Health, Roehampton, England, from Spare-Part Surgery *by Donald Longmore, Aldus Books, London, 1968.)*

Fig. 7–11.
Stages in the
assembly of an
artificial hand; the
finished product is in
the left foreground. The
hand in the right foreground
shows the motor and the lead screw
assembly that moves the fingers. The
thumb has two fixed positions and
can be set by the wearer in
one of two fixed positions
before the start of any
given movement.
(United Kingdom
Atomic Energy
Authority)

Why does it make sense to speak of goals in describing the activities of living things, whereas it is nonsense to speak in the same way of inanimate physical objects such as the falling rock?

Perhaps you may conclude, as some people have done, that goal-seeking behavior is the unique characteristic of life. But is this always so? An elevator operator has the goal of bringing the elevator to rest at each floor so that the floor of the elevator cab is level with the floor of the building. In many older elevators, the speed of the elevator is controlled directly by the operator. At the end of the day when the elevator is heavily loaded with people on their way home, the operator must begin to bring the elevator to a stop before it reaches the desired level if he is not to overshoot his mark. If he does overshoot, he will reverse the motion until the floor of the elevator is at approximately the same level as the outside floor. With practice, a skilled operator can learn to adjust his actions in accord with the number of passengers. He stops the elevator at the desired level without a number of adjustments.

Modern elevators now have automatic mechanisms for accomplishing the same goal much more effectively. The mechanism built into the automatic elevator remembers at which floors it should stop to pick up passengers and at which floors to discharge passengers. It can also stop at the proper level more consistently than it could under the control of a human operator. The automatic elevator, thus, can be said to exhibit a goal-seeking behavior, despite the fact that it is completely inanimate. What then is the essential difference between the elevator and the falling rock that lets us describe one, but not the other, as a goal-seeking device?

The goal-seeking behavior of the elevator is made possible by the presence of a *feedback* arrangement in the elevator system.

The difference between *mechanistic behavior* (which is governed only by past causes) and *purposeful behavior* (which is guided by future or desired goals) has been a battleground of debate among philosophers, theologians, psychologists, and many others. Yet it is only within the past thirty years that man has discovered the concepts needed to understand some of the issues involved. The most important of these concepts lie at the heart of this course. Of them all, the central concept is that of *feedback*, which permits us to build machines and systems which display goal-seeking characteristics.

Box 7–1

GOAL-SEEKING DESCRIBED BY AN EQUATION
(*System of Fig. 7–12*)

If we call the input I and the output O, the error is

$$\text{Error} = I - O$$

The output is just K times this error, or

$$O = K(I - O)$$

This equation can now be solved for O by ordinary algebra:

$$O = \frac{K}{1 + K} I$$

The boxed equation describes the relation between the input I and the output O.

For example, if K is 4, the output is $\frac{4}{5}$ of the input. Regardless of how we change the input, the output is always $\frac{4}{5}$ as large.

When K is very large (e.g., 200,000), the output is practically equal to the input. In the elevator-control system, the output is the floor level of the car. We then use as the input the level of the floor at which we wish to stop. The feedback system brings the car floor to the desired level. The output automatically seeks the goal.

Fig. 7–12.
*A simple
feedback system.*

Fig. 7–13.
System
regulating the
price of shoes.

3 | FEEDBACK AS SELF-REGULATION

Feedback can be regarded as goal-seeking, and we can consider many feedback systems to be *automatic regulators*. These are systems which automatically try to hold the output at some desirable value.

Figure 7–13 shows the block diagram of a simple economic system. The output is the price of a certain product—for example, men's shoes of a particular quality. Suppose the normal or average price is $12. When the actual price in the stores is $14, the shoes are priced $2 above normal. The comparator output is then $2 (Fig. 7–13). This higher price causes more manufacturers to appear. People owning shoe factories start to make more shoes of this quality. The total quantity manufactured increases, causing an oversupply. The oversupply in the marketplace then causes the price to drop because the manufacturer must lower his price in order to sell his shoes. Now the price falls to $10—$2 below normal. The resulting lower profits cause some manufacturers to leave the business. The quantity manufactured falls, and the demand in the marketplace for the decreased supply results in higher prices. The price moves up toward $12. This is a remarkable system—it automatically adjusts the price to the normal $12. When the price is low, it automatically rises; when it is high, it automatically falls.

Figure 7–13 also illustrates one other factor which we often find in feedback systems: *instability*, or the tendency of the output to change out of control. The output may oscillate indefinitely— first high, then low, then high, and so on, or the output may simply grow and grow. This latter instability results because there is a loop within a feedback system. Signals can travel around and around the loop (Fig. 7–14). If we start with an error of $\frac{1}{100}$ and it is multiplied by 10 as it travels around the loop in one second,

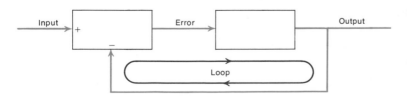

Fig. 7–14.
The loop
in a feedback
system.

the error is $\frac{1}{10}$ a second after the start. A second later it is one. After only eight seconds, the error is 1,000,000.

This tendency toward instability in feedback systems can be illustrated by familiar examples:

1. The relation between sleep and health is sometimes a feedback system. When you have a sore throat or nasal congestion, sleep is difficult. The lack of sleep or rest tends to make the cold worse, which in turn makes sleep even more difficult.

2. Economic systems frequently exhibit feedback. When a union as large as the auto workers obtains a sizeable wage increase, costs of automobiles tend to rise. Other costs rise as the added income of the union members results in more spending in other industries, hence greater demand for goods. The resulting increase in prices makes the union ask for further wage increases. A single union, of course, does not control this inflationary spiral, and the system tendency to oscillate can be controlled by changes in government spending and taxation.

3. The U.S.A.-U.S.S.R. armament race illustrates feedback. The United States learns of the development of the U.S.S.R. intercontinental ballistic missiles and launches a major program to develop more missiles than the Soviets possess. Learning of the United States missile arsenal, the U.S.S.R. then undertakes development of a major anti-missile system. In order to maintain the "balance of power," the United States must then launch an urgent program for an effective anti-missile system. Each of these steps is expensive for each country. In addition, each country is using valuable men in jobs which are not really productive.

4 | FEEDBACK FOR DISTURBANCE CONTROL

The examples of feedback in the last section were all cases in which feedback exists inherently in the system. In the armament race, feedback is inevitable since the United States' actions influence what the U.S.S.R. does. If feedback only occurred in such situations, and instability was the only consequence of feedback, there would be little point to our detailed discussion here.

Fortunately, we can intentionally use feedback to improve the way a system works. *Control over unwanted signals is a very important use of feedback.* Often unwanted signals affect our system output. These disturbances are unpredictable. They ruin system performance by causing undesirable output. When a system which includes feedback is operating properly, the output is continuously controlled. When the output changes because of signals beyond our control, corrections are made automatically.

Fig. 7–15.
Navigating
through a channel.

For example, as a Navy pilot attempts to land his plane on an aircraft carrier flight deck, the primary input signal is his location relative to the deck. As he watches the flight deck from the cockpit, he continually adjusts his controls. There are also two secondary or *disturbance signals:* the wind gusts acting on his plane and the forces the sea exerts on the carrier. If these disturbances are too strong or too violent, successful landing becomes a most difficult task.

The problem of landing on a carrier is particularly complicated because it involves all three dimensions in space. Figure 7–15 shows a similar two-dimensional problem. A ship is located 10 nautical miles south of a channel opening through a reef. The navigator sights the buoys or light signals marking the channel and naturally sets a heading due north. A heavy fog suddenly sets in which hides the buoys from the men on the ship. The ship steams on at 10 knots (nautical miles/hour) with a north heading. The navigator, having correctly estimated the distance to the channel as 10 nautical miles, assumes that the ship will pass the reef in one hour—as indeed it will if there is no disturbance signal affecting the position of the ship. But what happens if a current of two knots pushes the ship eastward? The northerly motion is unaffected, but in one hour the ship is 10 nautical miles north and 2 nautical miles east of its original position, or 1.5 nautical miles east of the edge of the reef. Tragedy ensues.

In order to understand the system, a block diagram is useful (Fig. 7–16). The input signal is the heading the navigator orders for the ship. A disturbance signal (the current) adds to this heading to yield a signal which is the actual heading assumed by the ship. This actual heading determines the system output, the position of the ship relative to the center of the channel.

*Fig. 7–16.
Block diagram
for navigation
problem.*

The tragic ending can be averted if we can use feedback to counteract the effect of the current. For example, perhaps every 15 minutes the fog lifts long enough for the navigator to complete another sighting on the channel markers. After 15 minutes, the ship is at A which is 7.5 nautical miles south and 0.5 nautical mile east of the channel center as in Fig. 7–17. The navigator now orders a heading toward the channel center. Fifteen minutes later the ship is at B where another sighting is taken and a new heading adopted. (We still assume the navigator is stubborn. He refuses to recognize that his past errors might be the result of a constant current, for which he could compensate by aiming to the west of the channel center.) If this calculation is continued, we find that the ship follows the path shown in the figure, with sightings taken at locations A, B, C, and D.*

The block diagram of the system with feedback is shown in Fig. 7–18. Every 15 minutes, the navigator sees the channel. He adjusts the heading according to the difference between where he is and where he would like to be. In this example, feedback occurs every 15 minutes. We can do even better if we can use feedback continuously (i.e., no fog exists, so the navigator can sight the channel markings and adjust the ship heading continuously). Then the path followed by the ship is shown in Fig. 7–19. Initially the ship moves off the desired course because of the current, but the continual corrections result in final passage through the channel without difficulty. In this case, the feedback in the system is continuously compensating for the effect of the disturbing signal (the current). This navigation example illustrates in rather general terms a most important use of feedback. *Feedback can be used to reduce the effects of disturbance signals.*

Random Disturbances

The interesting feature of this property of feedback is that the effects of the disturbances can be controlled even when these signals are random and unpredictable. In our example, the current always flows at the same speed. In such a case, we do not require much feedback. After the navigator determines his position at the 15-minute mark, he can calculate the current. He then heads for an appropriate location west of the channel center so that, by the time he reaches the reef line, the current will have returned the ship to the desired midpoint in the channel. Feedback is needed

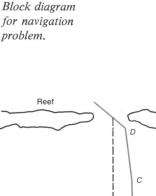

*Fig. 7–17.
Path followed
by ship with feedback
every 15 minutes.*

*
The path followed in Fig. 7–17 is most easily determined graphically if the diagram is drawn to scale. For example, once B is known, C can be found as follows. From B along a line toward channel center we mark off a distance of 2.5 nautical miles to determine a point which we can call C′. This is the point the ship would reach 15 minutes after leaving B if there were no current. C is 0.5 nautical mile due east of C′. Once C is known, D can be found similarly, etc.

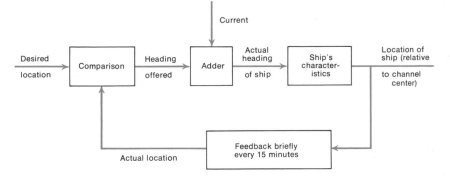

Fig. 7–18.
Block diagram
of system with intermittent
feedback.

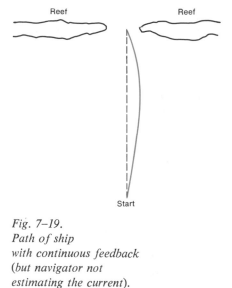

Fig. 7–19.
Path of ship
with continuous feedback
(*but navigator not
estimating the current*).

only once so the navigator can determine the speed and direction of the current.

This role of feedback is also illustrated by an auto steering system. As we drive down a traffic lane, we continually observe our position and make corrections, since the disturbances are unpredictable. The need for feedback is clear if we consider what would happen if we were blindfolded.

As you have read this section, hopefully you have participated in a systems problem. The author and publisher have combined to generate a small input to the student's mind. The system output is your level of understanding of the section.

Because your mind may possibly have been subjected to disturbance inputs as you were reading (interrupting telephone calls, television, and so forth), we should attempt to control the effects of such disturbances with the use of feedback. A simple form of feedback in this learning system depends on the use of a set of questions to measure the system output (your degree of understanding). If the output is not at the desired value, the questions refer you back to an earlier portion of the section so that the input signals can be repeated.

Question 1. What is a primary use of feedback? If you are at all uncertain of the answer, a complete re-reading of the section is recommended.

Question 2. When feedback is used, what signal must be measured? Once this measurement is made, feedback usually involves a comparison. Why? If you are uncertain, the navigation example should be re-read.

Question 3 (Box 7–2). In the block diagram of Fig. 7–22, the gain K is to be chosen so that the part of the output due to x is the same for the two systems. Determine K. Which system results in less influence of u on the output? By what factor is the disturbance effect decreased? If you have any difficulty with this problem, it would be advisable to return to the discussion of Box 7–2.

Box 7–2

EQUATIONS FOR DISTURBANCE CONTROL

We can describe disturbance control by algebra. Figure 7–20a shows a system with a disturbance input called *u*. The total output *y* is 20 times the main input plus 10 times the disturbance:

$$y = 20\,x + 10\,u$$

Figure 7–20b shows the same system (inside the dashed lines), but with feedback added with blocks *A* and *B* and the comparator. For this system, the output *y* is

$$y = 20[A(x - By)] + 10\,u$$

(20 times the input to the dashed box plus 10 times the disturbance.) Now if we use algebra to solve this equation for *y*, we obtain

$$y = \frac{20A}{1 + 20AB}\,x + \frac{10}{1 + 20AB}\,u$$

We can select *A* and *B* to give the desired behavior of the system. In our original system, we had $(20x + 10u)$. We might want the feedback to reduce the effect of disturbances by 40. Then we would like *y* to be $(20x + \frac{10}{40}\,u)$. Comparison of this with the last equation above shows that we want

$$1 + 20AB = 40 \qquad \text{(From the } u \text{ terms)}$$

$$\frac{A}{1 + 20AB} = 1 \qquad \text{(From the } x \text{ terms)}$$

Hence, *A* should be 40. Then we can find *B* from

$$1 + 20AB = 40 \qquad \text{when } A = 40$$

or

$$B = \tfrac{39}{800}$$

The final system is Fig. 7–21. By adding feedback, we can keep the *x*-to-*y* system unchanged, but reduce the effect of a disturbance by a factor of 40. Feedback permits control over unwanted signals.

We have now inserted the first type of feedback. Additional feedback is provided by your class discussions; with these various feedbacks, our theory tells us that it is certain that the student's level of uncertainty will be controlled regardless of any disturbance signals. When this system output is measured in the future by an exam, the change will surely be great.

This "example" itself emphasizes that the idea of feedback is important in people systems as well as machine systems.

$$y = 20x + 10u$$

(a) Original system with no feedback

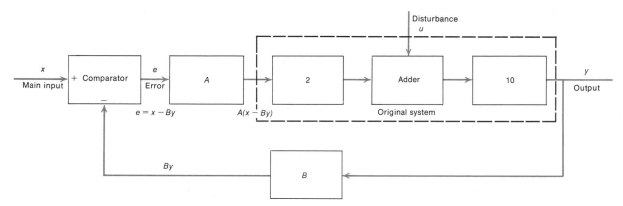

(b) System after feedback added

Fig. 7–20.
System to
describe disturbance control.

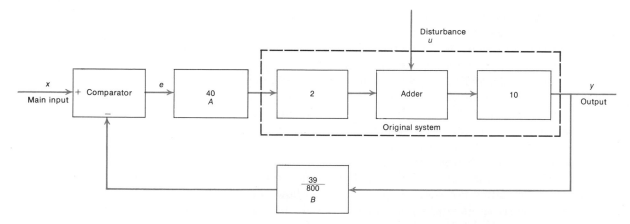

Fig. 7–21.
Final feedback system

$$y = 20x + \frac{10}{40}u$$

Fig. 7–22.
Two systems
with disturbance
input.

5 | AUTOMATIC COMPENSATION

Feedback has a second, very important advantage. *Feedback can correct automatically for changes in the system.* This characteristic of feedback is called the control of system sensitivity. How sensitive is the system to changes in its parts? Feedback can keep the senstivity small.

One of the most familiar examples of a feedback system is the regulation in the central heating system in a house. As we discussed in Chapter 5, a single furnace supplies heat to all parts of the house. The output of the system is the temperature. The input signal is the desired temperature for comfort. The thermostat measures the output and compares this measured value with the desired value (the input which is set manually on the thermostat). The system is shown in Fig. 7–23.

When the actual temperature drops below the desired temperature, the thermostat relay closes and the furnace is turned on. Heat then flows through the house. The actual temperature rise resulting from this heat flow depends on the thermal characteristics of the house (the windows and doors that are open, the insulation provided from the outside, and the degree to which air is circulating through the house). Regardless of these thermal characteristics, however, heat is supplied by the furnace until the actual temperature rises to about 3° above the desired temperature. At that point, the thermostat relay opens, the furnace is shut down, and the heat flow stops. The speed with which the changes occur depends on the particular form of the system. For example, if heat is transmitted through circulating hot water, heat continues to flow into the room from the radiators after the hot water stops circulating, until the water cools. If the desired temperature (the thermostat) is set at 72°F, the living room temperature near the thermostat fluctuates between about 70°F and 75°F and never exceeds these limits as long as the outside temperature is low and the furnace system is large enough to heat the house properly. The performance

Fig. 7–23.
Elements of
household heating
system.

of the system is *completely independent* of the characteristics of the house and the outside temperature.

By using feedback, we have a system which is nearly perfect. We can even hold the front door open in the winter for a long time without losing control of the inside temperature.

A second example of feedback controlling sensitivity is shown in Fig. 7–24. It is a picture of a mechanical amplifier to increase the effective strength of a man. The machine is worn by the operator like an external skeleton, attached at the feet, forearms, and waist. As the man moves his arms, the exact motions are repeated by the machine powered by hydraulic motors (similar to the power steering and power brakes used in cars). The device can lift loads of 1500 pounds.

In order to permit the man to control the machine despite widely varying loads, a fraction of the load forces acting on the skeleton are applied to the man's arms and legs. For example, if the machine's arm hits an object, the operator feels a fraction of the force on his own arm. In this way, the machine becomes an extension of the man. The operator can use the machine in a normal way to move unusually heavy or cumbersome loads or objects that are particularly dangerous, such as bombs. The force feedback to the human being is the key to successful operation of machines of this type.

As in the last section, we attempt to add feedback to the learning system. In this learning system which involves the author, publisher, and reader, the characteristics of all three elements vary widely; we should attempt to ensure success by adding feedback in the form of a few simple questions.

Question 1. What are the two primary uses of feedback in system design (this section and Section 4)?

Question 2 (based on Box 7–3). For the system shown in Fig. 7–26, determine the percentage change in overall system gain when the amplifier gain falls by 20%.

Question 3 (based on Box 7–3). Figure 7–27 shows a more complex feedback system. Determine the change in overall system gain when the amplifier gain decreases by 90%. This is a rather unusual system, since it turns out that y is independent of the amplifier gain even though y is the output of the amplifier. Thus, even if the amplifier gain dropped to 0.001, the output y would be unchanged.

Fig. 7–24.
*This is a
model of a set of
"mechanical muscles"
that will give a human being
the strength of a giant, and permit
him to lift a 1500-pound load while
exerting only a fraction of this force.
Attached to its operator at his feet,
forearms, and waist, the machine,
nicknamed HardiMan, will mimic
and amplify his movements.*
(*General Electric Company*)

Box 7–3

SENSITIVITY CONTROL WITH FEEDBACK

The ability of feedback to correct automatically for system changes is explained by Fig. 7–25. For the system without feedback we have

$$y = 10x$$

The output is ten times the input.

For the system with feedback

$$y = 1000 \left(x - \tfrac{y}{10}\right)$$

If we use algebra to solve for y, we have approximately

$$y = 10x$$

Thus the two systems behave in essentially the same overall fashion. Each gives an output which is ten times the input.

Very often the gain of an amplifier varies during operation. If the total amplifier gain drops by 10%, what happens in each system?

The system without feedback is easy. If the amplifier gain falls by 10% (from 10 to 9), the output $y = 9x$: the output also falls by 10%. The feedback system behaves quite differently. The total amplifier gain is now 900 (10% less than 1000), and

$$y = 900 \left(x - \tfrac{1}{10} y\right)$$

If we solve for y, we find

$$y \cong 10x$$

The output is unaffected by the 10% change in amplifier gain. We have rounded off the numbers here. There is actually a very slight reduction in system gain. Thus, feedback makes possible satisfactory system operation, even when the characteristics of system components change rather radically in the course of time.

The extent of the value of feedback is even more apparent if we ask the reverse question: In the feedback system above, how much must the total gain of the amplifiers drop from 1000 before the output drops by 10%? If the amplifier gain is G (instead of 1000), the output is

$$y = G\left(x - \tfrac{1}{10} y\right)$$

If y is to be equal to $9x$ (rather than the normal $10x$)

$$9x = G\left(x - \tfrac{9}{10} x\right)$$
$$9x = G \tfrac{1}{10} x$$
$$G = 90$$

The amplifier gain must drop from 1000 to 90 (more than 90%) before the feedback system loses 10% in gain.

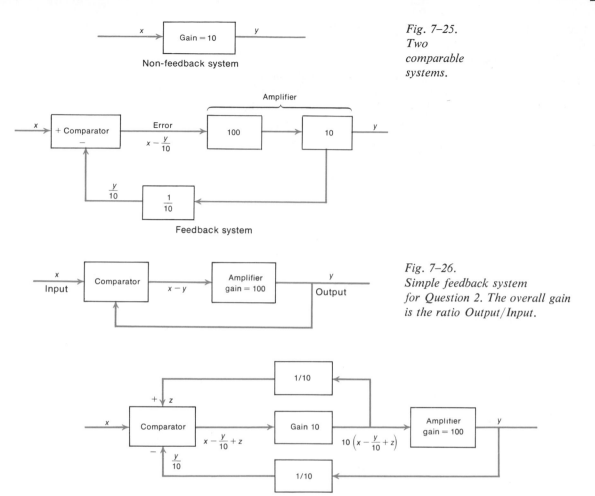

*Fig. 7–25.
Two
comparable
systems.*

*Fig. 7–26.
Simple feedback system
for Question 2. The overall gain
is the ratio Output/Input.*

The system suggests some of the amazing things that can be done with feedback.

In this and the preceding section, the two basic purposes of feedback are described. We next look at a familiar feedback system to see these properties.

*Fig. 7–27.
System
with two separate
feedback paths.*

6 | A HUMAN FEEDBACK SYSTEM

Consider the system in a human being for the control of the internal body temperature. The system illustrates the use of feedback to give very accurate system performance even when there are large disturbance signals or large system changes.

The normal temperature of the core of the body (the internal organs and the central nervous system) is about 98.6°F. (The British consider the normal temperature to be 98°; in this country we use 98.6° since this corresponds to exactly 37° Centigrade.)

This temperature must be very accurately controlled. Cells of the central nervous system are damaged if the temperature rises as much as 7°F. An even smaller drop in temperature results in greatly reduced enzyme activity within the body. In the normally healthy individual, the temperature does not vary more than 2°F either way.

The core temperature is held almost constant even when the body is exposed to outside temperature changes of more than 100°F. We recognize the similarity between this system and the example of the last section, the control of temperature inside a house. In the human system, however, nature provides several different means by which temperature is measured as well as several different sources of heat, rather than the single thermostat and furnace common in houses.

In order to construct a model of our system, we must understand the basic elements of the system. From a thermal viewpoint, the body consists of three principal parts: (1) the core, (2) the skeletal muscles, (3) the skin.

In the control of core temperature, the human being uses several different methods to vary the heat in the core:

Chemical. The basal metabolic rate (BMR) is generated primarily in the core. The oxygen inhaled is brought by the blood to the cells where fat is stored. The oxidation or burning of this fat results in a release of carbon dioxide to the blood as well as a release of energy. When the BMR is measured during a physical examination, the patient is not permitted to eat for at least 12 hours in advance of the examination and must have complete rest during that time. The net oxygen consumption of the individual is then measured to determine the BMR, the rate at which the man is internally converting fat to energy. Metabolic temperature control is achieved through an endocrine gland which receives electrical signals from the brain.

Shivering. The muscles also provide a source of heat. If the sensors in the skin detect a sharp drop in outside temperature, electrical signals are transmitted from the brain to the muscles to order shivering. Here adjacent muscles (the same ones normally used for motion or useful work) operate in an uncoordinated fashion, with the result that there is very little useful work and most of the work is converted to heat.

Skin changes. The skin is used to effect changes in internal temperature in two ways. First, the blood flow to the surface of the skin can be controlled (this is called the vasomotor effect). When heat flow out of the body is to be decreased, less of the warm blood goes to the skin. Second, sweating leads to heat loss by evaporation, and is particularly important when the man is in a very hot place.

Thus, the human being has four ways to control the internal body temperature: variation of the metabolic rate, shivering, changing blood flow to the skin, and sweating. Each method is controlled by that part of the brain which regulates body temperature.

We now have described in general terms how the system works. We are ready to make a block diagram. The output of our system is the body temperature (the temperature of the core). Another output signal (Fig. 7–28) is the skin temperature. In addition to the four primary input signals to the body, there are disturbance inputs: changes in heat flow when the man exercises or when the outside temperature changes.

The primary input to our system is the desired core temperature (98.6°F). In case of illness, this input is probably increased to produce a fever. (Medical research has not yet indicated how this change is accomplished.) In the brain, the error is measured by comparing the actual core temperature with 98.6. This error is then used to change the core temperature. It is interesting to note that the human system of Fig. 7–28 also includes a measurement of skin temperature, in order to anticipate heat demands when the external temperature changes rapidly (as it does when one enters a hot or cold shower or moves from the inside to the outside of a house during cold winter weather). In our analogy to the home-heating system, the skin sensors correspond to outdoor thermometers connected to the thermostatic control system to anticipate sharp changes in the outside temperature.

The block diagram of Fig. 7–28 includes the main signals which make up the temperature-control system of the human being. What is the value of a diagram of this sort? How does such a description of the system aid the researcher in learning about the operation of the system? How might the model help in developing improved medical procedures?

Fig. 7–28.
Thermal regulating system for temperature of body core.

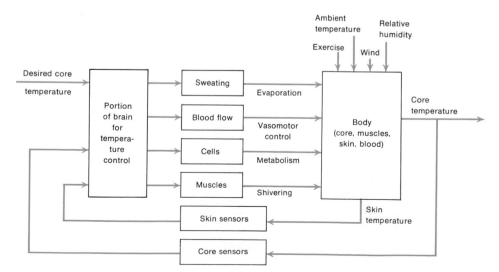

The answers to these questions are difficult unless we develop a much more detailed model. We should look at each block in Fig. 7–28. In each case, we would try to learn to describe the block mathematically. We might have to make measurements on men. We certainly would want to understand the physical laws which explain how things work. If we did this here, however, we would end up with a book on physiology, rather than a book on the man-made world. We can make certain comments just from the simple model.

First, the importance of feedback is apparent. The internal body temperature changes very little (typically less than a degree) when the man passes through radically different environments: feedback almost eliminates completely the effect of disturbance signals. Second, the core temperature is almost independent of changes in the body system. Major changes can occur (we might have a limb removed). Minor changes such as changing clothes are common. As a result of feedback, we have a system in which the performance is the same under all these conditions.

The model indicates basic questions, for example, how is the input changed (if it is) to cause a fever in a patient who is ill? Further study of this question may indicate that the input is not changed, but rather that the system loses effective control during illness. If the latter is true, perhaps major efforts should be made to hold down the temperature of an ill patient. On the other hand, maybe if we could learn more about the body, we would find that the fever is an important contributor to the body's power to resist disease. If so, we should not indiscriminately use drugs which reduce the temperature—at least not unless the body temperature approaches the danger range. When a child has a fever, the mother usually gives him an aspirin to bring the temperature down. Perhaps this is the wrong thing to do, at least if the temperature is not too high.

Furthermore, our model is only approximate. In the future, as we do more experiments, we will find new parts of the model. Then new experiments will be suggested. Hopefully, this process will lead to better and better understanding of the system. Thus, the feedback model is a key tool in the development of scientific understanding.

Finally, the model of this part of the human being may show us how to improve man-made systems. We saw that the human system measures skin temperature so it can anticipate large changes in the outside temperature. In the same way, we can build a better house-heating system if we measure outside temperature. When the outside temperature starts to fall, we could increase the heat flow inside to prepare for the cold wave.

Now let us consider a final feedback question. In hibernation what changes in the system would you expect? (The model might represent a woodchuck instead of a man.) What elements of the system would work during hibernation? Ordinarily, in hibernation the core temperature of the woodchuck drops to about 39°F. What advantages are there with this low temperature?

7 | INSTABILITY IN FEEDBACK SYSTEMS

Feedback is primarily of interest in goal-seeking systems. In the example of navigating through the reefs, the goal is to place the ship in the center of the channel. In the human temperature-control system, the goal is a constant core temperature of 98.6°F, regardless of surrounding temperatures or physical or emotional activity.

The success of feedback in controlling both disturbances and changes in the system involves certain disadvantages. We have already seen that feedback normally requires a more complicated system (the output signal must be measured and compared automatically to the desired value of the output in order to calculate an error signal which can be used to correct the output). In this section, we discuss a second, major disadvantage of feedback: *the possibility that the system may be unstable.*

In very general terms, a system is unstable if its output goes out of control. Perhaps the most dramatic example of instability is the hydrogen bomb; here a small detonation rapidly grows into an immense explosion. There are many other examples. Visitors to Bermuda observe the abundance of lizards on the island. Some years ago, a few lizards were brought to the island to control the mosquitoes; the lizards rapidly multiplied, the mosquitoes disappeared, and the islanders are now worrying about controlling the lizard population.

Perhaps a more familiar example is the instability occurring when a car goes from an understeer to an oversteer condition. Understeering means that as the car travels around a curve it tends to pull out of its curved path, or to increase its radius of curvature (Fig. 7–29). To follow the road, the driver must apply more and more force to turn the steering wheel. In the oversteer case, the automobile tends to decrease the radius, or to "tighten" the turn; to compensate, the driver must apply less force on the steering wheel during the turn.

The study of automobile behavior during turning is an exceedingly complicated problem since the motion depends on the road angle and surface and on the rapid actions of the driver. A dangerous situation occurs when an automobile moving at a certain speed passes from an understeer to an oversteer condition; the driver, if he is not aware of the change, tends to turn the steering wheel incorrectly to try to correct the car position. A problem in automobile design is to be sure that such a switch cannot occur suddenly and surprise the driver.*

* There were hundreds of law suits against General Motors in connection with the Corvair car manufactured during a few years in the early 1960's. These were based in large part on the claim that poor engineering resulted in such a transition from understeer to oversteer. It was claimed that the sudden change made the car impossible to drive and accidents resulted. In the first case brought to trial with outstanding engineers testifying for both sides on the technical aspects, the judge decided for General Motors. The engineering problem in such a case is so complex that scientific, mathematical, or computer analyses and experimental tests often yield no clear conclusions.

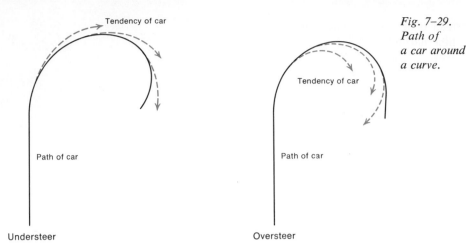

Fig. 7–29.
Path of
a car around
a curve.

Understeer   Oversteer

Example of Instability

As another example of instability, we can return to our system of Section 1 of this chapter: a man picking up a pencil. As we saw there, the man observes the difference between the actual output and the desired value. Corresponding electrical signals from his brain to the muscles produce the required change in the output (Fig. 7–30). In addition to the main feedback through the eye, there is an additional feedback path directly from the muscular system to the brain: as the muscles are actuated, the individual feels the forces and resulting motion.* As soon as the hand touches the tabletop or pencil, there is an additional feedback path (not shown in the figure) which results from the sense of touch.

The above feedback system is normally excellent. It operates satisfactorily even when the disturbing signals are present (e.g., motion of the system when the task is performed inside an airplane or other moving vehicle) or when system characteristics change (the individual is tired, physically weak, or distracted mentally by other events).

Instability may exist, however, when individuals suffering from an illness called ataxia try this task. The hand starts to oscillate as it approaches the pencil. The man is unable to pick up the pencil because of the violent shaking of his hand. The loss of control comes from oscillation or instability within the feedback system.

*
The existence of this internal feedback path is apparent if the man closes his eyes after observing the pencil location. He still can sense how far his hand has moved and approximately where his hand is located.

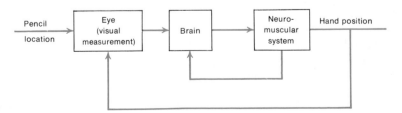

Fig. 7–30.
Feedback system
involved in picking up
a pencil.

Fig. 7–31.
Feedback system
with feedback path
delayed.

Instability and Delay

There is no simple way to find when instability occurs in a feedback system. We can understand some aspects of the problem from the simple system of Fig. 7–31. Here the input is compared with the output, and the error causes the output to change to reduce the error. Suppose now that there is a delay as shown. The output is measured, but not sent to the comparator until a short time later. If the time delay is such that this output reinforces (or adds to) the input, the system may become oscillatory: the input causes a certain output, this is fed back and adds to the input, the even larger output is in turn fed back, and the output signal grows larger and larger.

The above description is certainly not exact or any real help in determining whether a feedback system will be stable or unstable; it does indicate that delay in the system is a key to instability. The delay (which we find in every real system) is important to instability in a feedback system. In general, the greater the delay, the more tendency there is for the system to become unstable.

In the examples considered previously, the cause of the delay is obvious. In the system for positioning the hand to touch a pencil, the delay comes from the time required for the brain to reach a decision and the time needed to move the muscles.

An Economic System with Delay

In inventory control, a company observes its sales of a particular product and then decides on a manufacturing schedule. If the company wishes to keep its customers happy, it must have enough completed products in its warehouses to satisfy rush orders. On the other hand, completed and unsold products in the warehouse represent dollars not working for the company. Thus, decisions are needed which will establish the best policy.

The feedback system is complicated by two delays: the delay in manufacturing (if the demand for the product rises sharply, time is required to increase the factory output because of the delays in obtaining raw materials and the time required for manufacture), and the delay in measuring changes in orders and demand for the product. Instability (or loss of control) results in very large changes in factory activity (and the resulting costs from large changes in the number of people working, the training of new employees, and so on). Instability may lead to warehouses bulging with unsold products or to an inability to meet customer orders (with the dissatisfied customers then turning to a competitor's product).

Instability in Biology

The population growth of hydra is a biological example of instability. Hydra are very small, freshwater animals which increase rapidly in numbers when the food supply is plentiful. One class of hydra can be fed exclusively on water fleas. As the hydra population in a closed container grows, the individual animals tend to become smaller (an automatic adjustment over generations to the population explosion). This regulating or control system may be unstable: the hydra become so small they are unable to eat the water fleas, and the entire population dies of starvation. In this case, the feedback mechanism (decreasing individual size with population growth) leads to the end of the system.

Such an extreme effect of feedback in a biological system probably results from the artificial environment created in the laboratory. In a natural environment, food sources would be available in at least small quantity. Some hydra would survive, even though the total population might decrease. This decrease in population would then result in larger animals, which would in turn thrive on the water fleas, and the total system would tend to oscillate around an average population and size.

Thus, in biology, feedback seems to cause instability when the natural environment is changed by man. With the rapid increase in technology in recent years, we now seem to be able to change the environment in very important ways.* Will instability result? If instability results, will this instability be permanently harmful to the human race? Unfortunately, we can only answer such questions if we understand the feedback system in detail.

Useful Instability

Finally, instability can be useful. We can use instability to obtain a sinusoidal signal. In the pendulum shown in Fig. 7–32, the ball (hung on a string attached to the ceiling at O) is moved to A and released. The motion is familiar: the ball swings down to B, then on to C where it reaches a maximum height, then back to B and on toward A. As the ball oscillates back and forth, a little energy is lost each trip because of air friction and friction at the pivot. Consequently, each return to the left is a little lower than the preceding, and the pendulum finally comes to rest at point B. Figure 7–33 shows the way the position of the ball changes as time passes.

If we wish to keep the oscillation going indefinitely, we can give the ball a very slight push each time it returns to the extreme left. We do the same thing when we pump a swing to keep it

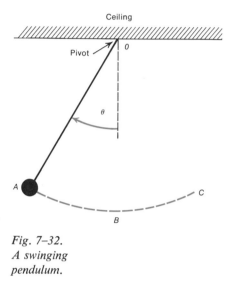

Fig. 7–32. A swinging pendulum.

*
The possibility of major success in weather control (particularly rainfall) is a particularly timely example. A more dramatic example is the current possibility of placing in orbit at an altitude of 22,500 miles (so that it sits over the same spot on earth) a large reflector to reflect sunlight back to the earth's surface during hours when the sun is below the horizon. The darkness of night would thereby be avoided—with the consequent, undetermined effects on nature as well as on man's living habits.

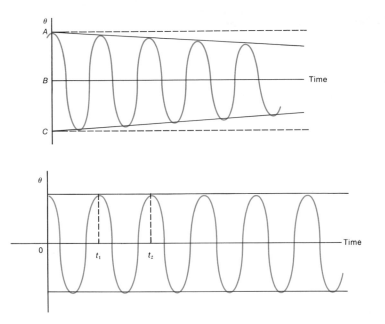

Fig. 7–33.
*Gradually
decaying oscillation
of a free pendulum.*

Fig. 7–34.
*Pendulum
oscillation with
regular excitation.*

moving. With this slight push in each cycle, the position signal
has the form shown in Fig. 7–34.

If this were a ball swinging from the ceiling, control by a human
being continually tapping the ball would be tiresome. Instead, we
can automate the system by surrounding the area at A by an
electromagnet, which is energized in such a way as to repel the ball
at the top of each swing (the ball would have to be magnetized
also, of course, because we are using the force between two mag-
nets). Thus, at instants t_1, t_2, and so forth in Fig. 7–34, a pulse of
current through the electromagnet gives the ball a slight push as
it starts its downward swing: just enough of a push to compensate
for the energy lost during the preceding cycle. If we wish, even the
timing of the electromagnet current can be controlled by the
motion of the ball so that the current starts slightly after the ball
enters the electromagnet near the end of the travel toward A. The
system, called an *oscillator*, is then entirely automatic and maintains constant-amplitude oscillation of the pendulum. The output
is a sinusoid, as we discussed in Chapter 6.

An oscillator system is the basis for an accurate clock operated
by a battery rather than by winding or from the electric power
lines. The clock requires no winding and operates for the life of
the battery. In this case feedback is used to yield an oscillation.
What we want is an output which is constantly changing. We are
really using feedback to obtain a controlled type of instability.

The main purpose of this section is to emphasize that feedback
may be associated with instability. The amount of feedback which
can be used in a system is often limited because we must have a

stable system; otherwise, control is meaningless. When we use feedback in any system (economic, social, or man-made), we must be aware that the system may become unstable.

8 | FINAL COMMENT

Feedback is a way of looking at system problems. To understand and to design complex systems, it is often convenient to think of the system in terms of a block diagram. In a block diagram, feedback is represented by measurement of the output and comparison of this measured value with the desired value. The error is then used to change the output to reduce the error.

Feedback has been used by engineers for centuries. Perhaps the earliest engineered feedback system was the plumbing device developed by the Romans and used in much the same form today, the water-level control in the tank at the back of the common toilet. Here a ball floats on the water as the water level rises. When the water level reaches the desired height, the rising ball shuts a valve which stops the incoming water. When the water in the tank is released by flushing, the tank empties, a rubber stopper then covers the outlet, and the cycle repeats.

The first major work on feedback-system engineering was carried out at Bell Telephone Laboratories during the 1920's as the telephone engineers developed a system for long-distance telephony. Telephone conversations were possible across the United States only if amplifiers were used every few miles to amplify the voice signals. Once a large number of amplifiers were included, however, satisfactory system performance required that the gain of each amplifier should not change very much; otherwise, the volume for the listener might vary greatly. To build amplifiers constant over long periods of time, the engineers utilized feedback.

Feedback engineering made another major advance during World War II, primarily because of the importance of feedback systems in aiming large guns and radar antennas. In earlier wars, guns were smaller and targets moved very slowly. In World War II, antiaircraft gunfire required rapid aiming, which demanded more force than a man could provide. Machines had to be used. Feedback was essential if these automatic systems were to be accurate.

Since World War II, feedback engineering has continued to be a key part of modern technology. *Automation* involves automatic feedback control of decision-making processes, whether in factories, in traffic control, in the control of anesthesia during operations, or in the navigation of space vehicles. In addition, during the past twenty years the idea of feedback has been used in the study of biological, social, and economic systems. Automation is perhaps the most important technical development in recent years, and feedback is the central idea in automation.

Fig. 7–35. Designs of experimental 300-mph trains. In both cases, the train rides on an air cushion. *High-pressure air is forced down from under the vehicle, so that the vehicle sits on this air rather than on the solid surface.*

One of the major problems is how much to bank such a train as it goes around a curve. The banking angle for maximum passenger comfort depends on the speed and the sharpness of the curve. In a feedback system, the speed and curvature are measured. Air coming from the four cushions underneath is automatically adjusted to give the proper bank angle. Feedback is essential if the system is to work. (Aviation Week and Space Technology)

Thus, it is appropriate to close this chapter with a description of an automated system. Figure 7–36 shows a diagram of the Alfred E. Perlman automated freight yard of the Penn-Central railroad in Albany. When a freight train arrives at the yard from a distant point, it includes individual freight cars which must be sent to many different locations. In other words, the incoming trains must be broken up and the cars reassembled into outgoing trains bound for different destinations.

There are places for 7973 cars within the yard; on the average 3000 cars are switched each day. There is room for cleaning 100 cars at a time and repairing 113 others. There are 70 parallel tracks in the classification yard where cars are sorted and new trains assembled.

The entire system is controlled by a central digital computer. Each car is pushed by a locomotive over a hump 26 feet high. As the car then rolls down the incline, it is identified and its speed and weight measured. By the time it reaches the bottom of the incline, switches are automatically thrown to route the car onto a track where a train for its destination is being assembled.

As Fig. 7–36 shows, the speed of the car is controlled automatically to bring it smoothly into contact with the rest of its train, and the weather conditions are even fed into the digital computer since they influence the rate at which the car will slow down when it reaches the level. In addition to all the feedback instrumentation and computer control, the entire operation is monitored by six different television systems, 12 radio systems, and 15 teletype machines to communicate information to the men supervising this process. If the railroads are to provide low-cost, high-quality service, feedback is essential.

Fig. 7–36.
Humping operation in automated railroad freight yard.

Questions for Study and Discussion

1. Your dog sees another dog and dashes off to battle. If you whistle to him and he obeys you, would you call this "goal-seeking behavior"? Give your argument in favor of your answer.

2. "Plants extend their roots toward moist and fertile soil." "Cells at the surface of the root differentiate to form . . . long extensions of the wall, the *root hairs*." Discuss the process which leads to the result stated in the first sentence above. Show how a feedback mechanism is involved.

3. An escalator or moving stairway must run at a speed which is as nearly constant as possible, whether nobody is on it or it is crowded with passengers.

   a) How can such control be accomplished?

   b) What is the input? The output?

   c) Illustrate with a block diagram.

4. In each of the following feedback systems tell what is the input signal and the output signal.

a) A machine in a sugar refinery to fill boxes each with just 2 lb of granulated sugar.

b) An automatic record player: the device to return the pickup arm after finishing a record.

c) An electric refrigerator.

5. Describe briefly at least one feedback mechanism in each of the examples given in Question 4.

6. The temperature control of a shower is an interesting feedback system.

a) Construct a block diagram for the system assuming that the man standing in the water spray attempts to control the temperature of the water which reaches him by adjusting the hot-water control knob only. The knob is so placed that he can reach it while standing in the spray.

b) The simple block diagram which you have probably derived can be changed to include the fact that when the man changes the knob setting, he estimates the time delay before the temperature change will be detectable at the surface of his skin. Show this new block diagram if time delay was not included in your answer to (a).

c) The statement is made in the text that time delay is often associated with instability. Explain briefly how instability is likely to result if the man underestimates the time delay between the turning of the knob and an observable change in water temperature.

d) The stability problem of a feedback system is often complicated by parts of the system in which a change in input results in no change in output over significant ranges. This is one example of what the mathematician and the engineer call nonlinear behavior. For example, in many shower systems there is appreciable *backlash* in the knob: as the knob is turned clockwise, the amount of hot water decreases; but if we try to reverse direction (and move counterclockwise), the first 15° or so of motion result in no change in hot water flow. It is only after we have turned the knob through 15° that control is reestablished. This "slippage," or backlash, occurs every time the direction is reversed. Describe briefly how such a common mal-

function may make effective temperature control more difficult (and may actually lead to instability).

7. In each of the following situations, decide whether feedback is or is not involved, and briefly explain your reasoning. (In each case, somebody had to take some action, and he, of course, is a bundle of feedback loops. Leave him out of the discussions.)

a) A golfer uses a 7-iron for an approach shot. His ball acquires backspin and hence does not roll off the green.

b) A weight is hung from a support by means of a spring. It is then pulled down an inch or two and set free, after which it bounces up and down for a long time but finally comes to rest.

c) A billiard ball rolls against the rail of the table and bounces back toward another ball.

d) A coin is dropped in the slot of a vending machine. The door opens and allows the buyer to pull out a bottle of soft drink.

8. In manufacturing thin sheets of metal, like aluminum foil, it is important to keep the thickness constant, but equally important not to stop the rolling machine to gauge the thickness. A mechanical device for continuous measurement would probably mar the sheet, even if it were able to make precise enough readings. The solution adopted is to place a radioactive source of alpha particles (a bit of the element plutonium is often used) at one side of the sheet, and a detector for alpha particles at the other. The rate at which the particles get through to the detector is exactly dependent on the thickness and material of the sheet. Show how such a gauge could be made part of a feedback loop to control the thickness of the sheet (use a block diagram if you prefer).

9. A famous medieval philospher, Jean Buridan, told of a donkey placed between a bale of hay and a bucket of water. His hunger and his thirst were so exactly balanced that he starved to death because he could not decide which desire to satisfy first. Nowadays we might imagine that an oscillation resulted, with the consequence that he worked

himself to death. Briefly explain how such an oscillation might be established.

10. Suppose in the pencil problem described in the first section, the man was blindfolded but the pencil was tapped loudly as it was placed on the table. Draw and explain the block diagram which represents this situation.

11. On more than one occasion an engineer has constructed an "automatic animal." The beast has motor-driven wheels and carries a battery. If it runs into an obstruction it stops, backs off, turns a little, and tries again. When its battery runs low, it rolls to its hutch, brightly lighted inside, and there plugs itself into a battery charger until it is powered up for more exploration of its environment. In general terms only, how can these accomplishments be built into the "animal"? Can you think of a way to equip it so that it could run to the edge of a precipice, stop, and go away from the precipice?

12. Describe the function in maintaining constant body temperature of:

a) the BMR (basal metabolic rate).

b) the muscles.

c) the skin.

13. Draw a block diagram which describes the feedback involved in an armament race between two countries.

14. Draw a block diagram which describes the feedback during a political campaign.

15. Describe the feedback system which you would propose to improve education in America, your state, and your school system, or your classroom.

Stability

8

I | INTRODUCTION

When we design systems to cope with the world in which we live, we usually have in mind some desired operating state or condition. This desired state may be very simple: we may wish to place a book on a table so that it remains stationary. Or the desired state may be quite complicated: we may wish to place an astronaut into a precisely calculated orbit.

Sometimes, describing the operating state may be easy: we want the book to remain in a fixed and unchanging position. At other times the description of the operating state may be difficult: we must compute the astronaut's precise orbit with the aid of high-speed computers. But in either case, the design must involve additional information. We must know if the operating state is *stable* or *unstable*. An unstable design may produce disastrous effects. A man in a satellite sent into an *unstable* orbit would be catastrophic. His distance from the earth would either increase or decrease until tragedy occurred.

In this chapter we will probe into the concept of stability. We will find that unstable and stable systems differ in behavior if a disturbance is applied to them when they are in equilibrium. In many systems, we can deal with the phenomenon of stability in a quantitative manner by applying our techniques of modeling and analysis. By applying these techniques we can determine when a system changes from a stable to an unstable state.

2 | SKYSCRAPERS BEGET SKYSCRAPERS

There is a common saying, "skyscrapers beget skyscrapers." What does this mean? Until the beginning of this century, office buildings were generally a few stories in height (often only three or four stories). During the early 1900's, skyscrapers began to appear in several of our major cities. The trend toward multi-story buildings rapidly accelerated until today a large percentage of urban office buildings are at least 30 stories high.

The advantage of a tall building is that we can centralize many operations. Workers can live near their place of business, thus cutting down transportation costs. With taller buildings, land can be used more efficiently. Buildings such as the Empire State Building, more than 100 stories high, occupy only about $\frac{1}{100}$ of the land that would be needed for all the tenants if they were to be arranged on a single floor.

Some disadvantages, however, are associated with the height. As a building is made taller, the foundations must be made sturdier to support the added weight. With the all-masonry buildings that existed at the end of the last century, gigantic foundations were required for tall buildings. Elevators were non-existent, and people simply would not or could not walk up more than a few flights of stairs.

A natural competitive state of equilibrium existed. Every office building owner realized approximately the same profit, and all office buildings in a large city were generally about the same in height. The profit equilibrium was stable. Any attempt to disturb this equilibrium with the construction of a taller building proved wasteful since no one would walk up many flights of stairs. Hence the income necessary to pay for the additional cost of the foundations could not be found. A building of fewer stories did not realize the income of the buildings of the normal height.

But toward the end of the century two changes occurred. First, the cheap fabrication of steel beams made possible tall, lightweight, steel-frame buildings. Second, Otis invented the elevator, and electrical motors became available for their operation. The equilibrium state for three- or four-story buildings could have continued to exist because the new inventions could have been ignored. But such equilibrium was now obviously unstable.

The first skyscraper brought its owner large profits from a small plot of land. This produced an increase in the value of the land on which the skyscraper was built. Owners of vacant land were no longer willing to sell at the price that prevailed when buildings were only a few stories in height. They could wait for a skyscraper builder who would be willing to pay more for the same plot because his building would realize a larger return for his investment. With increasing land prices, only a skyscraper was profitable. Each skyscraper encouraged new, and for a time, higher skyscrapers. In New York, for example, the instability was especially severe. (In addition, the bedrock of Manhattan made skyscraper building easy.) Recently, in cities such as London where the foundation conditions for tall buildings are much less favorable than in New York, the "skyscraper begetting skyscraper" effect has become quite noticeable.

The preceding example really describes two different systems. The first system is one that existed during the last century. This was a stable system: if anyone tried to build a skyscraper, he found it unprofitable, and buildings were held to three or four stories.

With the invention of the elevator and the availability of low-cost steel, the system was changed; it was now unstable. After the first skyscrapers were built, all construction had to be skyscrapers. Land became too valuable to permit low buildings. Today in city after city we find low buildings being torn down and replaced by skyscrapers. Even apartment houses are frequently 20 stories high. Urban universities are characterized by a set of structures towering above the campus.

Thus, discoveries in technology changed our urban construction system from stability to instability. In the new system of the twentieth century, a new operating point had to be found in order to achieve stability once again.

Two features associated with the stability or the instability of a system are illustrated by our example. These are:

1. In a stable system, a disturbance which moves the system will lead to forces that tend to return the system to the original state.
2. In an unstable system, a disturbance will cause forces that tend to drive the system even further from the original state.

Again and again, we meet these characteristics: the tendency of a stable system to restore itself and the tendency of an unstable ystem to "snowball" away from the equilibrium. A knowledge of these characteristics is essential for the creation and control of many systems in the man-made world. Although a qualitative appreciation of these characteristics is enough in many instances, in others we need a quantitative mastery: a knowledge of the numerical values of the quantities that may change the system from a stable to an unstable state. Later in this chapter, we apply modeling and analysis to some systems in order to illustrate the method of approach to such problems.

3 | STABILITY IN TRAFFIC FLOW

The snowballing effect in unstable situations is illustrated by a variety of examples taken from our normal life. Some of the most frustrating examples are provided by automobile traffic in a city. Anyone who has driven from the center of a city to the suburbs during the evening rush hour is familiar with the sharp transition from instability to stability in this system. Typically, as we move away from the city, traffic is bumper-to-bumper and moving at a very low average speed (perhaps 10 miles/hour in badly congested situations, even on a freeway or expressway). This slow rate of movement, with frequent stops, continues for some distance, even though a few cars leave the freeway at each exit as the individual

Fig. 8–1.
A common type of model used in the planning of traffic facilities in a city. This is called a "desire-line diagram." To construct such a diagram, we mark the start and finish of each trip. These two points are then connected by a line. Thus, the thickness of a solid mass of lines represents the number of trips along this direction. (Traffic in Towns, *Her Majesty's Stationery Office, London, United Kingdom*)

drivers turn off for their homes. There comes a time, however, when a few cars leave an exit and, as a result, the average traffic speed rises rapidly, perhaps to 50 miles/hour. The departure of just a small percentage of the total number of cars changes us from an unstable system to a stable system.*

In this example, the location at which the system becomes stable depends not only on the traffic density, but also on such factors as the road condition, the weather, and the location of accidents which block certain lanes of the expressway. An accident or stalled car may cause extreme instability, but the stable situation after we have passed the accident may actually result in a lower total time to reach home. Ice on the road during winter months causes radical changes in the driving habits of motorists and, hence, major changes in the conditions for stability.

The mathematical determination of what traffic density will cause instability is difficult. The model includes not only the physical characteristics of the highway, but also the manner in which drivers behave under different conditions such as how closely

*On the author's road home from work, this transition usually occurs at about the same point. A few miles further on, the expressway narrows from six lanes to four (two each way); the system is then again unstable, and it is another ten miles before stability returns.

they follow the car in front, and the extent to which they weave in and out of lanes. The problem is so complicated that it is usually easier to determine the characteristics of a given highway by *measurements* of automobile speed as a function of traffic density, time of day, weather, road conditions, and so forth.

Once such measurements are made, we have a model which permits the traffic engineer to determine the maximum number of cars which can be on a given stretch of highway if the system is to be stable (and, for example, average speed is to be at least 40 miles/hour). This model is the basis for modern systems for control of traffic flow on major city expressways. The number of cars on the controlled stretch of road is measured automatically by one of the following means:

1. Pressure transducers in the pavement and actuated by a car going over them.
2. Current-carrying loops of wire over the road. When a car passes under the loop, the magnetic field is changed because of the steel in the car. This change in magnetic field is detected electrically.
3. Photoelectric transducers (electric eyes). A light beam shines across the road on a device which generates electricity when illuminated. The passage of a car breaks the light beam and hence can be measured electrically.
4. Radar.

Distributor roads
Access roads
Housing
Areas of pedestrian freedom
Main pedestrian routes

Bus stop
School
Shops

Fig. 8–2.
One city-planning approach to the reduction of traffic congestion. This is known as the Radburn plan, since it was first used in Radburn, N. J., in 1928. In this approach, a central section exists free of vehicular traffic. In addition, there are pedestrian paths entirely separate from vehicular roads, so that people walking are protected from traffic and vice versa. (Traffic in Towns, *Her Majesty's Stationery Office, London, United Kingdom*)

This information on the number of cars on our controlled stretch of highway is sent to a computer in which are stored the measured model and also the current data on weather, time of day, and location of accidents or stalled cars. The computer then transmits appropriate control signals to the traffic lights at each entrance to the highway. When the automobile density is approaching the point of instability, cars trying to enter the highway are held back by a red light. The duration of the stoplight is controlled by the computer signals. The closer the system is to instability, the longer the delay for motorists waiting to enter the highway.

Limitations of the System

The primary limitation of such a traffic-control system comes from the need to win public acceptance. When a traffic-control system was installed on a portion of the Eisenhower Expressway in Chicago, the waiting period for each motorist was controlled up to a maximum of 14 seconds. If there are longer waiting periods (and of course there may be few cars within sight on the expressway ahead), the motorist tends to go through the red light or complain about the ridiculous traffic engineers. In general in a democratic form of government, it is exceedingly difficult to enforce laws which the public finds unreasonable.

This unwillingness of the motorist to accept control exists even though his own personal welfare is best served by obeying the control. Successful traffic control is even more difficult to achieve when, as commonly happens, an entire group of motorists are best served if everyone obeys the regulations, but any one motorist can disobey and benefit if the others obey. The temptation is then strong for the individual to think only of his own welfare. Once enough people are selfish, the traffic-control system collapses.

Thus, a basic limitation of an expressway traffic-control system arises from a lack of public understanding of the technology.

A second major limitation of such an expressway control system is the interaction of this system with other parts of the total automobile transportation system of the city. Any given expressway can clearly be kept stable if we prohibit any cars entering and enforce the regulation with enough policemen. Indeed, stability would be ideal if we kept all cars off the highway. An immediate result, however, is to clog the parallel highways and the streets feeding into the expressway.

Thus, it is not possible to treat only a small part of the overall highway system. We cannot break the complex system down into a large number of simple systems, since rapid and smooth traffic flow in one portion often means extraordinary congestion in the neighboring portions.

This interlocking of all parts of the total system is actually much worse in most cities of Europe than in the United States.

Most American cities have grids of north-south and east-west streets. Consequently, one-way streets can be used, and motorists can easily seek alternate parallel routes.

In the typical European city, and to some extent also in Boston, the streets tend to radiate outward from the center and to meander in various directions which have evolved over centuries from the old footpaths and horse trails. Furthermore, streets in Europe are usually much narrower than streets in the United States. Finally the construction of expressways within European cities is often impossible because of the scattered historical landmarks which must not be destroyed. When an expressway interchange may require 60 acres or more of cleared land, there are few parts of downtown London where such highways could be built.

As a result of these factors, urban traffic congestion in many European cities is already extremely serious, even though the percentage of the population owning cars is still far below that in the United States. The nations of western Europe are now approaching the period when most of the population can afford automobiles. As a result, it appears that some of these cities will within a few years display instabilities more severe than any encountered so far in the United States, particularly since road construction is already so far behind the current needs.

In this section, we have described in general terms a system in which stability is a most important goal. The model in the case of our expressway control must be measured by observation of actual performance of the system. The conditions under which instability occurs are determined by this observation. Fortunately, there are also problems in which we can work with a mathematical model in order to detect instability.

4 | THE BLACK DEATH

One third of the people in the world dead, 36% of the universities closed, 50% of the clergy dead, Europe devastated in only three years. Towns deserted, farms lying idle, and gloom over the entire continent. This is the picture in 1351 at the end of the Black Death, the bubonic plague that swept from Asia throughout the civilized world in the great epidemic of the fourteenth century.

Some of the most interesting stability studies are those which describe the spread of epidemics. In this section, we take a brief look at the bubonic plague epidemic, the Black Death. This major catastrophe illustrates the kind of instability modern epidemiologists attempt to anticipate and prevent by the use of models.

Bubonic Plague

Bubonic plague is a disease spread primarily by fleas which usually live on rodents. The disease is spread among the rodents (often rats) by the fleas, and humans contract the disease from fleas, especially when the rat population decreases because of disease. The fleas then seek human hosts.

Actually, the bubonic plague of the fourteenth century had three different forms. The first, involving bubons, or boils, at the lymph nodes in the groin or armpits, is spread primarily by fleas. In the second form, in which pneumonia is involved with disease of the lungs, transmission can occur directly between human beings. The third form, involving the nervous system, is highly contagious and causes death within hours.

The first recorded instance of the plague occurred in the Near East in the eleventh century B.C. Apparently, there are regions of northern India and central Asia where the plague is continually latent. Worldwide epidemics have started there at least three times during the last 2000 years.

The first great pandemic (an epidemic covering the civilized world) occurred in the sixth century A.D. Lasting fifty years, "Justinian's plague" involved the Roman world and certainly was one factor in the disintegration of the Roman Empire.

The latest pandemic started in the middle of the last century in China, reached Hong Kong by the start of this century, and then spread throughout the world from the Chinese ports. With an estimated 10 million deaths in India, the epidemic was most serious in Asia. With the concept of quarantine established in Europe by about 1720, the emphasis on sanitation and hospitalization during the last century, and improved urban sewage-disposal and rat-control systems, the epidemic did not seriously affect the western world. In more recent years, particularly with the use of antibiotics, bubonic plague has been well controlled with only 200 deaths per year over the last decade.

The great pandemic of the fourteenth century started in northern India. It is not known why the plague began its spread. Apparently the rodent population which normally carries the disease was forced to migrate, because of food shortage, excessive rain, earthquakes, or other reasons.

The plague reached western Europe in 1347 and primary effects lasted for the next three years during which more than 30 million Europeans died, or a third of the population. Actually the epidemic followed the normal pattern and recurred over the next two centuries before finally subsiding, but the term "Black Death" usually refers to the first catastrophic three years.

While the plague undoubtedly was brought to Europe in several ways, historians often spotlight the primary source as the Tartar siege of the Crimean port of Kaffa in 1347. As so often happens in war,* thousands of men were brought together in unsanitary conditions, in this case in a region infested by rats. The Tartars began dying by the thousands† and finally dispersed.

---

*
Until this century, war casualties were primarily from disease rather than battle. In the Napoleonic wars, for example, the British lost 26,000 in action, but 194,000 from disease.

†
They actually catapulted corpses into the city in the hope of infecting the defenders, although such an action was totally unnecessary.

ATLANTIC

OCEAN

NORWAY

SWEDEN

RUSSIAN STATES

DENMARK

SCOTLAND

IRELAND

ENGLAND

DECEMBER 1350

TEUTONIC ORDER

LITHUANIA

POLAND

JUNE 1350

HOLY

DECEMBER 1349

ROMAN

JUNE 1349

FRANCE

EMPIRE

DECEMBER 1348

HUNGARY

WALLACHIA

BULGARIA

AQUITAINE

REPUBLIC OF VENICE

JUNE 1348

SERBIAN PRINCES

NAVARRE

PAPAL STATES

CORSICA

KINGDOM OF NAPLES

ALBANIA

ARAGON

PORTUGAL

CASTILE

BALEARIC ISLANDS

SARDINIA

DECEMBER 1347

GRANADA

KINGDOM OF SICILY

MEDITERRANEAN   SEA

MOSLEM   STATES

Fig. 8–3.
Approximate chronology of the Black Death's rapid sweep through Europe in the middle of the 14th century is indicated on this map, which shows the political divisions as they existed at the time. The plague, which was apparently brought from Asia by ships, gained a European foothold in the Mediterranean in 1347; during the succeeding three years only a few small areas escaped. (From "The Black Death," William L. Langer. Copyright © 1964 by Scientific American, Inc. All rights reserved.)

Once the siege was lifted, ships immediately sailed for Genoa, Italy. When the ships reached Genoa, they were sent away and then carried the plague to Sicily, Spain, and northern Africa.

The chronology of the epidemic gives a picture of the plague fanning out from the Italy-Sicily start northward, westward, and eastward:

1348   N. Africa, Spain, France
1349   Austria, Hungary, Switzerland, Netherlands, northern Germany, England
1350   Scandinavia, Scotland, Ireland

In any one location, the response curve followed a common pattern of growth to a peak, then a gradual fall-off as a significant fraction of the population had developed immunity. (For example, Fig. 8–4 for the city of London.)

The size of the catastrophe is emphasized by the death rates, which varied from $\frac{1}{8}$ to $\frac{2}{3}$. Investigations of the death toll in Europe at that time are reported by W. Langer in the February, 1964, issue of *Scientific American:*

THOUSANDS OF DEATHS PER WEEK

TIME—WEEKS

Fig. 8–4.
Death rate
figures for London during
the Black Plague.

As reported by chroniclers of the time, the mortality figures were so incredibly high that modern scholars long regarded them with skepticism. Recent detailed and rigorously conducted analyses indicate, however, that many of the reports were substantially correct. It is now generally accepted that at least a quarter of the European population was wiped out in the first epidemic of 1348 through 1350, and that in the next fifty years the total mortality rose to more than a third of the population. The incidence of the disease and the mortality rate varied, of course, from place to place. Florence was reduced in population from 90,000 to 15,000; Hamburg apparently lost almost two-thirds of its inhabitants.

Langer further reports that

It is now estimated that the total population of England fell from about 3.8 million to 2.1 million in the period from 1348 to 1373. In France, where the loss of life was increased by the Hundred Years War, the fall in population was even more precipitate. In western and central Europe as a whole the mortality was so great that it took nearly two centuries for the population level of 1348 to be regained.

In many areas, the population did not return to its pre-plague level until 500 years later. The tragic extent of the epidemic was largely a result of the totally inadequate state of medical knowledge. The only treatments were bleeding, or the smelling or eating of aromatic herbs and unusual foods.

Popular Response

The response of most of the people to the plague was strongly fatalistic, in accord with the historical era. People believed that the disease was spread by elements in the air and, hence, moved according to the winds. There was widespread acceptance of the idea that the plague was God's sentence on a sinful population and, hence, that there was no reason to attempt to fight it.

As a consequence of this attitude, some groups of people (particularly in central Europe) became flagellants, publicly beating themselves. The hope was that God would be satisfied with this self-inflicted punishment.

A more tragic consequence of the ignorance of the way in which the disease spread was the widespread belief, particularly in France and Germany, that the Jews had poisoned the drinking water. The resulting mass burnings and massacres (16,000 in Strasbourg, 12,000 in Mainz, and so on) were not rivaled in barbarity in history again until the 1930's. Sixty large and 150 smaller Jewish communities were exterminated, with major shifts in the Jewish population (for example, to Poland and Lithuania).

Effects of the Plague

Because of the scarcity of carefully prepared histories of the fourteenth century, it is impossible to determine in detail the effects of the disaster. Clearly, the most profound aftermath was the deep impression the epidemic left on the people. The disease often dragged its victims through days of intense suffering, accompanied by extensive vomiting and a pervading stench surrounding the victim. When the bubons broke, black blood was discharged and the patient often died lying in a pool of the liquid. With an average of two people per household affected, few individuals in Europe escaped without direct contact with the plague.

In economic and social terms, the resulting shortage of farm labor tended to break down the power of the landowners and certainly was a factor in such social upheavals as the Peasants' Revolt in England in 1381. The tendency toward overpopulation in 1300 (at least in terms of farmland under cultivation) was abruptly reversed, wages of farm workers rapidly doubled, and long-term economic changes were initiated.

There was a particularly heavy toll among priests and monks, with mortalities as high as 50 percent. In part, this resulted from the fact that clergymen typically lived near graveyards; also, the monasteries were particularly tight communities in which the disease tended to hit everyone. As a consequence of the high mortality rate, new priests were ordained with essentially no training and selectivity, and there was a marked decline in the intellectual activity among the clergy which had provided such important leadership during the Dark Ages. Historians have argued whether the plague can be cited as a cause of the Reformation. Regardless of this question, however, the inability of the Church to protect the people from the plague certainly led to a diminished influence over the lives of the populace.

From the viewpoint of *The Man-Made World*, the primary interest in the Black Death is the way in which the spread of the plague conforms to the simple epidemic model. The study of past epidemics is the basis for refinement of such models so that the quantitative picture can be used to guide public health officials in future crises. For example, in the fall and early winter of 1968, an epidemic of Asian flu swept across the United States. Because of the problems associated with isolating the virus and developing a suitable serum, vaccine supplies were very limited and had to be restricted to the aged and people particularly susceptible. Vaccine was allocated to regions according to the model estimates of when the flu would strike particular locations, in order to optimize the effectiveness of the limited supply. In such a case, an accurate model for prediction is essential if treatment is to be effective.

Unfortunately, in the 1968 example the predictions based on epidemic models were not very good. In New York City, for example, the prediction was that the epidemic would peak about January. Vaccine was scheduled for late December. Actually, the peak occurred in mid-December and by January the epidemic had almost passed.

5 | EPIDEMIC MODEL

Some years ago a person suffering from smallpox was accidently permitted to enter the United States. To prevent the outbreak of an epidemic, an immediate search was launched to locate all persons who had been in contact with this individual since his arrival. The possibility of a rapid spread of the disease from an extremely limited beginning was well understood. The need to quarantine or isolate infected individuals to prevent widespread infection is really a process to prevent an unstable situation from developing. A single infection acts as a disturbance of the entire community which is originally in equilibrium. The city of Aberdeen in Scotland, for example, was affected by a typhoid fever epidemic in 1964. Strict measures were taken to quarantine the entire city to prevent a further spread of the disease.

This suggests that an epidemic has the mark of an instability phenomenon, with its sudden and abrupt change from one state to another caused by a disturbance. In the smallpox case, the disturbance was a single infected person. In the typhoid fever case, the initial disturbance was traced to an imported shipment of tainted fish.

For a number of years public health officials have developed mathematical models of epidemics in order to obtain deeper insight into the epidemic mechanism. For example, if many people in a community are immune to smallpox by vaccination, what would happen if a single infected person were to appear? He may infect some of the others. But if these in turn come in contact with only immune persons, no epidemic will develop. On the other hand, if each infected person infects three or four others, the epidemic will grow exponentially.

Another factor is that people who recover become immune. If they recover quickly enough, they may not infect enough additional people to expand the epidemic. How important are these factors relative to one another in the growth of an epidemic? Should the public health officials concentrate their resources on quarantine or on vaccination?

Such are the questions that one wishes to answer with the help of a model. In many situations, answers are difficult to secure because the phenomena are complex and influenced by unpredictable events. Nevertheless, the insights that the models have provided have been useful and have spurred further research.

We consider only the simplest models of an epidemic. Let us first analyze the following, somewhat artificial, situation. We assume that in a total population of N persons, everyone is either infected with a certain disease or is susceptible to catching the disease from those who have been infected. This model thus omits the possibility of individual recovery and subsequent immunity, death, or isolation. This is an oversimplification for most diseases, although for certain mild upper respiratory infections with a long period of infection, the model does have validity.

We denote the number of infected persons by i and the number of susceptible persons by s. We are interested in the rate of increase in i, that is how many new people will become infected each day. Let us denote the rate at which new people become infected by R_i. We expect this rate to depend on the number of persons who are already infected. The greater this number, the greater the probability that they will transfer the infection to others. But we can also expect that the number of newly infected persons on any day will depend on the number who are susceptible. Again, the greater the number of susceptible persons, the greater the number of people who will be infected.

Hence a reasonable model should have the number of newly infected increasing both with the number of presently infected, i, and the number who are susceptible, s. But what is the precise form of this dependence? We could, for example, assume that a certain fraction (1%) of the infected and 2% of the susceptible will produce newly infected persons each day.

Would such an assumption make sense? As a first, and simple, test of whether this model is useful or not, we might see what happens when i is zero. If we have no infected, s is just N (the total population). The model then says that 2% of the total population would predictably develop the disease despite the complete absence of carriers. Clearly the proposed model of the epidemic is not realistic.

Since the above model is apparently inadequate, we consider other ways in which R_i might increase with both i and s. Hence, we assume that the number of newly infected per day is some fraction of the *product* of the number of infected and the number of susceptible. If we assume that the fraction is 1%, the number of newly infected each day is given by the equation or model

$$R_i = 0.01si$$

This equation or model is free of the objections of the initial model. If there are no infected persons in the population ($i = 0$), there will be no newly infected. Likewise, if there are no susceptible ($s = 0$), there will be no newly infected. We may assume that the population consists *only* of infected or susceptible people so that $N = i + s$. Our new model of an epidemic is in fact generally used, although it is a very crude mathematical model. With this model we can now study the spread of the contagious disease and predict the nature of the epidemic.

This model can be represented in the form of a block diagram (or analog computer program) as shown in Fig. 8–5. We require the scalor, subtractor, adder, and integrator we used in earlier chapters, and also, one additional element, the multiplier for which the output is the product of the two inputs (in our case the output is si).

This block diagram appears rather formidable at first glance, but we can grasp its meaning with a step-by-step examination of its structure. We should remember that we are searching for the day-by-day value of the number of infected persons. We want to predict the growth of the epidemic and discover whether the peak will occur before conditions in the community become dangerous.

If the total population is N and the number of persons who have already been infected is i, then the number of people who remain to be infected is $N - i$. This represents the number of people who are susceptible, according to our model, and is implemented by the subtractor of the block diagram.

Because the value of i (the number of infected people on any day) is exactly what the analog computer produces as its output, we bring the output back to the subtractor. The output of the subtractor is then a numerical value for the number of people s who are susceptible to the disease.

Our mathematical model for the epidemic informs us that the rate of increase involves the product of s with i and the factor .01. We therefore tap the line which supplies the value of i and feed this value into a multiplier along with the value s. This output from the multiplier represents the product si. Multiplication in

Fig. 8–5. Block diagram of epidemic model.

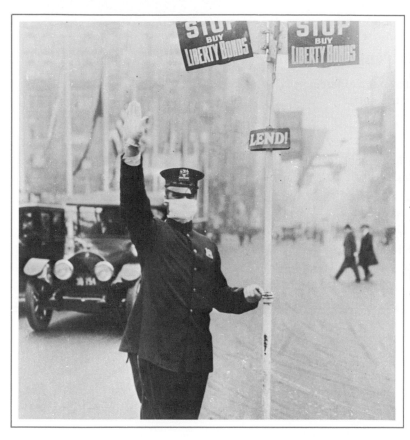

Fig. 8–6.
Signaling a virus to stop.
The mask worn by this New York policeman directing traffic at the height of the 1918 influenza epidemic offered little protection from the disease, for the tiny viruses that cause it could easily filter through. The 1918 scourge ranks with the Black Death as one of the most destructive holocausts ever to sweep the earth; in a few months it took some 20 million lives. (U. S. War Dept. General Staff photo no. 165-WW-269G-7 in the National Archives.)

the scalor unit by .01 produces the output .01si or the desired value of R_i.

The factor R_i is the daily rate of increase in the number of infected cases. If we send this information into an integrator, we can sum up the total number of newly infected cases after any period of time. This is the change in i (that is, Δi) which, when added to the initial value i_o, gives the quantity i or:

$$i = i_o + \Delta i$$

A meter attached to the output of the system and read at intervals of time can be used to graph a curve which simulates the course of the epidemic.

A Numerical Solution

The block diagram of Fig. 8–5 can be used in the laboratory for experimental determination of the growth of the epidemic. Alternatively, we can use the diagram as a guide for numerical calculations, if we want to study a specific example. If we carry through in succession each step indicated by the block diagram, we can find the way in which i changes each day.

Box 8–1

NUMERICAL STUDY OF FIRST EPIDEMIC MODEL

The block diagram (Fig. 8–5) shows the different relations involved in the model. As a specific example,

$$N = 100 \qquad i_o = 1$$

During the First Day

$s = N - i = 100 - 1 = 99$ We first calculate the number susceptible (the subtractor).

$si = 99 \times 1 = 99$ The output of the multiplier is si.

$R_i = \frac{1}{100}si = 0.99$ The rate of increase of infected is found.

$\Delta i = R_i \times 1 = 1$ One additional person is infected during the first day.

$i_1 = i_o + \Delta i = 2$ At the end of the first day, two people are infected.

During the Second Day

$s = N - i = 100 - 2 = 98$ And we go through the entire procedure again to find i_2, the number infected at the end of the second day.

The calculations of the box can be summarized in tabular form.

During day number	i at start of day	$s = 100 - i$	si	$R_i = .01si$	$\Delta i = R_i(\Delta t) = R_i(1)$	i at end of the day
1	1	99	99	0.99	≈ 1	2

Then, on the second day, the number of infected (new i) will be the number from the first day (old i) plus the newly infected (Δi) or $1 + 1 = 2$. Hence, on the second day, we can repeat the calculation. A continuation of this calculation gives Table 8–1.

During day number	i at start of day	$s = 100 - i$	si	$R_i = .01\ si$	$\Delta i = (R_i)(1)$	i at end of day
1	1	99	99	0.99	1	2
2	2	98	196	1.96	2	4
3	4	96	384	3.84	4	8
4	8	92	736	7.36	7	15
5	15	85	1275	12.75	13	28
6	28	72	2016	20.16	20	48
7	48	52	2496	24.96	25	73
8	73	27	1971	19.71	20	93
9	93	7	651	6.51	7	100
10	100	0	—	—	—	—

Table 8–1. Epidemic calculation.

Figure 8–7 shows the epidemic of Table 8–1 in graphical form.

The desirable state of a population is, of course, one in which no disease exists. We see from Table 8–1 or Fig. 8–7 that, according to our model, this desired state is unstable. A disturbance in the form of one infected person sets off a chain reaction which increases the movement away from the stable condition. In our example, this movement from the initial stable conditions continues until the 7th and 8th days, after which the rate at which people become infected begins to decrease as we approach the situation in which the entire population is infected.

Thus, the equations or the equivalent block diagram make up our simple model for the growth of an epidemic. From the model, we can construct numerically (in a step-by-step manner) a solution for any given initial conditions. The block diagram is particularly useful if we want to build an analog computer to find the response experimentally. (This would be especially desirable if we wished to determine i for many different values of N, of the initial i, and of the coefficient of si in the equation for R_i.) In addition, the block diagram describes pictorially how to organize the calculations if we wish to determine i for a single example, as in Fig. 8–7.

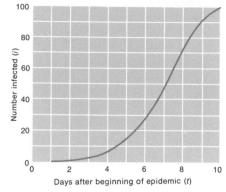

Fig. 8–7.
Plot of
simple epidemic model
given in Table 8–1.

6 | AN IMPROVED EPIDEMIC MODEL

The preceding model is just too simple to be realistic. We know Fig. 8–7 really makes no sense. By the end of the epidemic, there certainly will be some people who have not caught the disease.

One obvious problem with our model is that no one ever recovers, or even dies. Everyone who catches the disease stays infected. Can we model an actual epidemic if we allow for recoveries? How do we change our model to permit recoveries? A recovered person is certainly not infected or susceptible. Hence, we need a third category within our population. The population N then is the sum of three parts:

- s susceptible (people who can become infected)
- i infected (who have the disease)
- r recovered (or immune)

The next question is how many people have recovered? How does the number r change each day? Here we have several choices. We might say the number recovering each day equals the number infected ten days earlier, if the disease lasts ten days. The duration of the disease varies from person to person, and we might just assume some fraction (say 10%) of those infected recover each day.

With these assumptions about the recovered group r added to our assumptions of the last section, we can now find an "improved" model. Let us try to build up the block diagram step-by-step.

272

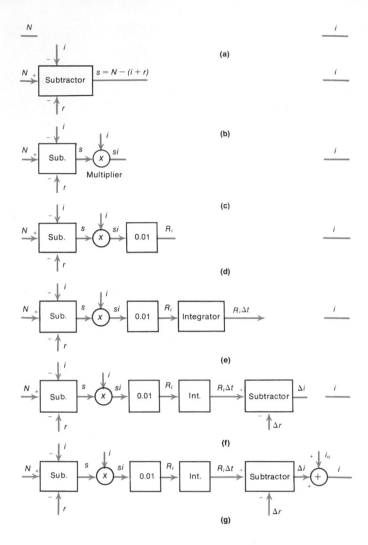

Fig. 8–8.
Block
diagram for improved
epidemic model.

Building the Block Diagram

Figure 8–8 shows the 11 steps in drawing the block diagram for our "improved" epidemic model. The following paragraphs are lettered to correspond to the parts of the figure.

a) Just as in the last section, the population N is the input. The output we hope to find is the number of infected i.

b) We first find the susceptible s (the total population minus the infected and recovered). Here we assume the signals i and r are available. We will have to find them later.

c) We need the product si.

d) The rate at which i is changing is $0.01si$, just as in the last section.

Fig. 8–8. (cont.)

(h)

(i)

(j)

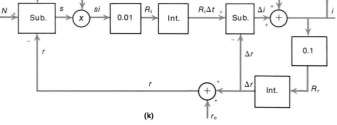

(k)

e) The increase in infected is the integration of R_i if we consider only those catching the disease.

f) When we include people recovering, the actual increase in i is the number newly infected $(R_i \Delta t)$ minus the number recovered Δr.

g) We now have found the total i on any day. The block diagram is complete except we have used r and Δr. We must now find what these quantities are.

h) We assume the recovery rate R_r is 10% of the infected.

i) The change in r is just the integral of R_r.

j) The actual r is the starting value (r_o) of those immune at the beginning plus the change.

k) The block diagram is now complete. This figure represents all details of our model. Notice that the block diagram is moderately complex. Actually, this is quite a simple system. In the modeling of an airplane wing vibration, housing growth in a city, or water pollution problems, it is not unusual to have several hundred integrators (instead of our two). Then the block diagram would be perhaps a hundred times as complicated.

In spite of the complexity of Fig. 8–8k, the construction of the block diagram is a step-by-step procedure. We work from the input to the output. Then we go back and find the new signals (r and Δr) which we introduced.

Numerical Calculation

Once we have the block diagram, we can calculate the behavior of an epidemic. The block diagram shows we need three different input signals:

N the total population
i_o the number of infected at the start
r_o the starting number of recovered, isolated, and immune people

Table 8–2 shows the results of the calculation when we start with 51,000 people, 1000 of whom are infected and none recovered.

Day	i	s	r	$R_i = 0.01is$	$R_r = 0.1i$	Δi Infections	Δr Recovered
1	1	50	0	.500	.100	.400	.100
2	1.400	49.500	.100	.693	.140	.553	.140
3	1.953	48.807	.240	.953	.195	.758	.195
4	2.711	47.854	.435	1.297	.271	1.026	.271
5	3.737	46.557	.706	1.740	.374	1.366	.374
6	5.103	44.817	1.080	2.287	.510	1.777	.510
8	9.118	39.604	2.278	3.611	.912	2.699	.912
10	14.888	31.740	4.372	4.725	1.489	3.236	1.489
12	21.208	22.119	7.673	4.691	2.121	2.570	2.121
14	25.544	13.284	12.172	3.393	2.554	.839	2.554
16	26.355	7.281	17.364	1.919	2.636	−.717	2.636
18	24.449	3.987	22.564	.975	2.445	−1.470	2.445
20	21.373	2.320	27.307	.496	2.137	−1.641	2.137
22	18.119	1.464	31.417	.265	1.812	−1.547	1.812
24	15.114	1.000	34.886	.151	1.511	−1.360	1.511
26	12.496	.732	37.772	.091	1.25	−1.159	1.250
28	10.276	.568	40.156	.058	1.028	−.970	1.028
30	8.422	.463	42.115	.039	.842	−.803	.842
32	6.889	.392	43.719	.027	.689	−.662	.689
34	5.627	.342	45.031	.019	.563	−.544	.563

Table 8–2. *Improved epidemic calculation.*

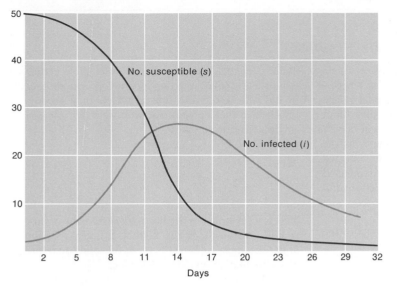

Fig. 8–9.
Plot of
epidemic given by
Table 8–2.

The numbers given are all in thousands (so we round off at the third number after the decimal point). We find the table by calculating the first row from left to right, then going to the second row, etc. Notice that at the dashed line we start giving results every other day. This is merely to shorten the table; we had to calculate values for each day before we could move on to the next day.

Plots of i and s are shown in Fig. 8–9. If the desired state is a disease-free population, then, in accordance with this refined model, such a state is unstable. One infected person triggers a "snowballing" away from the desired state. But our model predicts that not all of the susceptible people will develop this disease. The epidemic dies out before everyone is infected. This "improved" epidemic model, allowing recovered people, shows behavior which is similar to an actual epidemic.

Final Comments

The equation derived in Box 8–2 for the critical number of susceptibles quantitatively relates three factors that we know play important roles in an epidemic. Reduction in the number of susceptibles s to a value less than the critical value prevents the growth of an epidemic. This often suggests the need for vaccination and immunization. In some South American countries, for example, the malaria-preventative drug atabrine is added to the table salt in small quantities. Although not everyone uses table salt and different people use it in different amounts, it is hoped that enough immunity will be given to reduce s below the critical value. The rate at which the susceptible become infected, represented by the factor f_i, can be kept down by keeping susceptible

Box 8–2

MATHEMATICAL CONDITION FOR STABILITY

We know that the introduction of a single infected person into a population need not necessarily trigger an epidemic. Does our model reveal this possibility? What does our model predict about the conditions under which an epidemic will occur?

To explore these questions, we define f_i as the fraction of the product (si) that become infected in a single day and f_r as the fraction of infected persons who recover in a single day. (In our example, $f_i = 0.01$ and $f_r = 0.1$.) If these numbers f_i and f_r are changed, when is the system stable?

We can write the equations of the model in the form:

Change in recovered, $\Delta r = f_r i \Delta t$
Change in infections,
$$\Delta i = (f_i si - f_r i)\Delta t = (f_i s - f_r)i\Delta t$$

We note that the sign of the righthand side of the first equation is always positive. This indicates that the number of persons who have recovered will always increase. On the other hand, the sign of the second equation varies with the sign of the term in parentheses ($f_i s - f_r$).

If the number of susceptible persons in the population is large enough, ($f_i s - f_r$) is positive, and an increase in the number of infected persons results. However, if the number of susceptibles is small, the quantity ($f_i s - f_r$) is negative: an epidemic will not be initiated by the addition of another infected person.

According to our model, the borderline between instability and stability, that is, the condition which separates the epidemic from the non-epidemic response to an infection, is that in which:

$$f_i s_{cr} - f_r = 0$$

where we denote the critical number of susceptibles by s_{cr}. For a large number of susceptible people, the rate at which people become infected will be greater than the rate at which people recover and become immune or die. If the number of susceptible people is sufficiently small, the rate at which people recover or die will exceed the rate at which they become infected, so that no epidemic will occur.

Solution for s_{cr} gives

$$s_{cr} = \frac{f_r}{f_i}$$

This is a relationship for the *critical* number of susceptibles in terms of the factors f_i and f_r. In the above example, $f_i = 0.01$, $f_r = 0.1$ and therefore $s_{cr} = 10$, and an epidemic develops.

The importance of the values f_i and f_r is apparent if we change f_r to 0.6 and leave f_i at its original 0.01. Then

$$s_{cr} = \frac{0.6}{0.01} = 60$$

With an initial value of s of 50, the system is stable. We can compute the outcome of this condition in terms of the following table:

i	s	r	$f_i si$	$f_r i$	Δi	Δr
1	50	0	.500	.600	-.100	.600
.900	49.5	.600	.446	.540	-.094	.540
.806	49.054	1.140	.395	.484	-.089	.484

No epidemic results.

We note from the equation

$$\Delta i = (f_i s - f_r)\, i\Delta t$$

that the model has a self-limiting property. Even if s initially exceeds s_{cr}, as i increases we find that s must decrease. Eventually the factor ($f_i s - f_r$) becomes negative, after which the number of infected begins to decrease. If this occurs with sufficient speed only a small proportion of those susceptible to the disease will become infected.

persons away from contact with the infected ones through quarantine and isolation. Finally the rate at which people recover, f_r, is also important. We can expect to find that diseases which are contagious for long periods are more difficult to control than those with short periods of contagion.

We have considered only simple models of epidemics in this section. In more refined models, the laws of chance are included, such as various probabilities that an infected person will come in contact with another who is susceptible.

7 | LAW OF SUPPLY AND DEMAND

It is a commonly accepted truism that the "supply always adjusts to meet the demand." For example, when medical evidence recently indicated that it was healthy to be slim, demand for weight reduction drugs increased. In response to this demand, the drug manufacturers increased the supply of such drugs.

The "law of supply and demand" clearly involves the concept of a static operating state. In the example of the weight reduction drug, we may imagine the existence of a number of "demanders" who consume the drug at the same rate at which it is supplied by the manufacturers. If for some reason the demand for the drug were to decrease, we believe that the supply would also decrease until the attainment of a new equilibrium state.

But as we have already seen, it is also necessary to know about the *stability* of the operating state. In economic systems, as in all other situations involving people, this is a very complex question and it is influenced by many factors. Complicated mathematical models have been devised to deal with problems such as the supply-and-demand drug example. For simplicity of treatment, we focus attention on a small economic system consisting of a single supplier and a single buyer.

We consider the following situation. Mr. I. M. Crafty is a toy manufacturer. Mr. Crafty makes one toy and his principal outlet is Discount Stores, Inc., a chain that depends on high-volume sales. To cut costs, Discount Stores stocks as little as possible in warehouses, where rental and deterioration add to expense. Discount's policy is to place orders every Tuesday for an amount that can be sold in a single week. Their estimate of the weekly sales volume is based on the current price quoted by the manufacturers.

Mr. Crafty on the other hand sets his price on the basis of the most recent order from Discount Stores. Every Wednesday he sends letters to his customers announcing the new price.

What will happen? There is obviously a possible operating state where Discount orders the same amount Tuesday after Tuesday and Mr. Crafty sets the same price Wednesday after Wednesday; but will this operating state remain stable?

In order to consider the stability question, we consider a possible change in the normal pattern of operation. For example, Discount might decrease its toy order. Mr. Crafty then may increase his price in an attempt to maintain his total profit. Discount may feel that it cannot sell as many toys at the higher price and may further decrease its purchase order. If this action by Discount causes a further increase in price by Crafty, it is apparent that the system is unstable.

We might start out assuming a different disturbance in the normal operation of the system. Discount may place an even larger order than normal to take advantage of the high-volume price discount. Mr. Crafty could then increase his price because his toy is now in greater demand, which may result in a cut in the size of the order. In this case, the system seems to be stable. Clearly the situation is not simple; a multitude of possibilities exists.

A Quantitative Model

How can we apply modeling and analysis to gain an insight into this complicated affair? We can first try to model Discount's strategy of buying. Probably Discount will buy a greater amount the lower the price set by Crafty. If we plot price versus the amount that Discount will buy, the simplest curve of this sort would be a straight line as shown in Fig. 8–10. Here, for a given price P, Discount will buy an amount A.

Let us now examine and model one of Mr. Crafty's possible selling strategies. Recognizing that Discount Stores has increased its purchases because his price to them is low enough to gain a good profit on the resale of the toys, Crafty may attempt to induce a larger sale to Discount by a further reduction in price. The larger his sales to Discount, the larger the profit he hopes to earn. A simple straight-line model of this strategy is displayed in Fig. 8–11.

Figures 8–10 and 8–11 represent the quantitative model for the system. In this particular case, the two straight lines indicate the rules of behavior assumed for the two parts of the system: the manufacturer and the buyer. Obviously, in an actual problem the

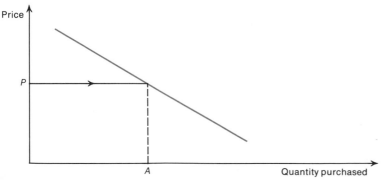

Fig. 8–10. Discount Stores' buying strategy.

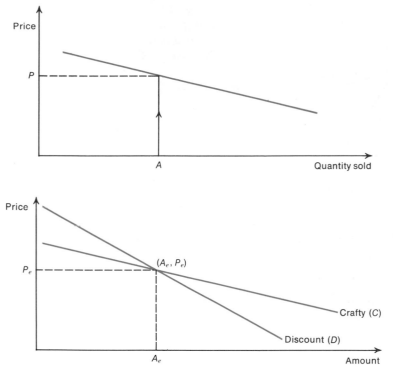

*Fig. 8–11.
Crafty's
selling
strategy.*

system analyst would have to determine each of these strategies by study of the operation of the two organizations; we merely assume these curves in order to illustrate the determination of stability.

Stability Analysis

On the basis of our straight-line models of Crafty's "supply policy" and Discount's "demand policy," what will happen? When will equilibrium of supply and demand occur? We can easily find this and also analyze for stability graphically. In Fig. 8–12 we plot both Crafty's selling strategy, labeled C, and Discount's buying strategy, labeled D. Equilibrium occurs at the intersection point A_e, P_e. At this price P_e, Crafty is willing to supply amount A_e week after week, and Discount is willing to buy this amount at the stated price. For a smaller purchase, Crafty insists on a higher price. Discount also expects to pay a higher price for a lower amount, but not necessarily the price Crafty expects. Likewise for an amount greater than A_e, Crafty's and Discount's expectations differ. Only for an amount A_e are they both in agreement on the price.

The point (A_e, P_e) represents what we call an *equilibrium point* or *equilibrium condition* (this is the reason for the subscript e). The system behaves properly as long as its operation rests at this

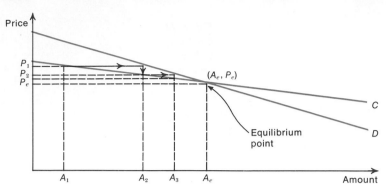

Fig. 8–13.
Stability of
supply-demand point for
Discount and Crafty.

point—as long as Crafty sets the price P_e and Discount continues to order A_e.

Stability of the system depends on what happens if a change or disturbance is introduced which momentarily moves operation away from equilibrium. For example, a fire in the toy department of one of Discount's stores compels them to reduce their order to A_1 in Fig. 8–13 (or a new buyer for Discount might underestimate the number of toys his firm could sell and reduces his order).

Then Crafty sets his price at P_1 in the same figure. At price P_1 Discount is willing to buy an amount A_2, as shown on the demand model in the graph. For this larger purchase A_2, Crafty offers a lower price P_2, which encourages Discount to increase its purchase further. This produces a cycle of adjustments until the point A_e is reached.

In other words, the system is stable. (We assume that both parties maintain constant strategies—the straight lines do not change.) After a disturbance initiated by Discount's low order of A_1, the system gradually returns to the equilibrium point. Indeed, we can show that for the strategies of Fig. 8–12, regardless of the nature of the momentary disturbance away from equilibrium, the system always returns to (A_e, P_e).

An Unstable System

On the other hand, if the strategies are those shown in Fig. 8–14, where the Crafty curve C is steeper than Discount's curve D, the system is unstable. If Discount now places an order for an amount A_1, which is less than the equilibrium purchase A_e, the amount purchased decreases; then Crafty increases his price. At the new price, Discount reduces his order. This cycle is repeated until Crafty has priced himself out of business. If in any one week Discount were to increase its order above A_e, the prices will continue to drop until Crafty loses all profits. The equilibrium is unstable in this case, as indicated by the arrows shown in the figure.

Examination of Figs. 8–13 and 8–14 shows that the stability criterion is a very simple one in terms of our model. For stability, *Discount's buying curve must be steeper than Crafty's selling curve.*

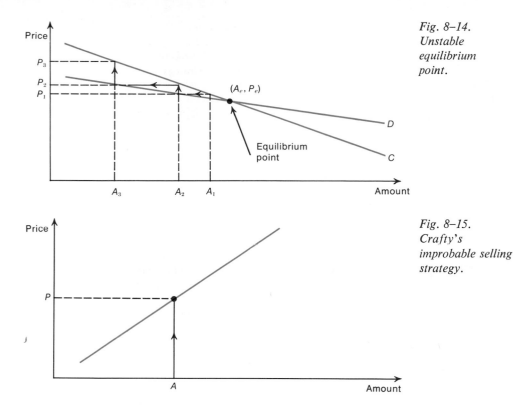

Fig. 8–14.
Unstable
equilibrium
point.

Fig. 8–15.
Crafty's
improbable selling
strategy.

Stability with Other Strategies

The preceding stability criterion or requirement is particularly simple; we need only determine the relative values of the slopes of the two strategy lines. The determination of stability is slightly different if we consider other strategies for either part of the system (either Crafty or Discount). For example, Crafty may believe that when Discount increases his purchase, the toy sells early during the week and is apparently in great demand. He may, in fact, ask a higher price. This rather unusual selling strategy is shown in Fig. 8–15.

Likewise, Discount may observe that their customers have peculiar buying habits: the customers believe that when an item is high priced, the item must be of high quality and therefore they buy at an increased rate. This buying curve is shown in Fig. 8–16.

Let us consider the stability of equilibrium when either Crafty or Discount adopts an unlikely strategy. For example, if Crafty adopts his improbable strategy, we have the case shown in Fig. 8–17. In this case a reduction to A_1 of Discount's order causes a cyclic pattern of buying and selling. The astute customer in Discount's store notices that for some reason the price of Crafty's toy fluctuates between P_1 and P_2 every two weeks. This cycle is in fact another equilibrium state, in addition to the point A_e, P_e.

Fig. 8–16.
*Discount
Stores' unlikely
buying strategy.*

Fig. 8–17.
*Supply-demand
when Crafty adopts
his unlikely strategy,
and Discount adopts his
likely strategy. (The
slopes have the same
magnitude.)*

Fig. 8–18.
*Stable supply-demand
when Crafty adopts his
unlikely strategy, and Discount
adopts his likely strategy.
(Slope magnitude for* D
is greater.)

In Fig. 8–17, the slopes of the C and D lines are *exactly* the *same* in magnitude, although the slopes have opposite signs (C positive and D negative). The result is the cyclic behavior shown. If the initial disturbance or change is smaller (nearer A_e), the cycle is smaller as well.

If the magnitude of the slope of the D curve is slightly greater than that of the C curve, then the situation shown in Fig. 8–18 occurs. (The equal slope case is shown in dotted lines for comparison.) Figure 8–18 indicates that point A_e, P_e is stable.

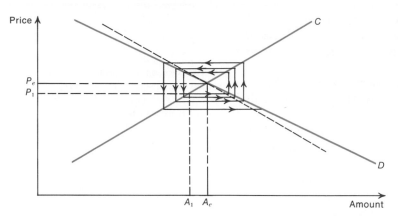

Fig. 8–19.
Unstable supply-demand
when Crafty adopts his unlikely
strategy. (Slope magnitude of D *is*
less than the slope magnitude of
C.)

Figure 8–19 demonstrates that if the magnitude of the slope of the *D* curve is less than that of the *C* curve, the equilibrium point A_e, P_e is *unstable*.

We have thus considered two special cases. Two other possibilities remain: one in which Discount uses his unlikely strategy (Fig. 8–16) while Crafty uses his likely strategy, or both Discount and Crafty go to their unlikely strategies.

With a graphical analysis of the two remaining possibilities similar to those outlined above, it can be shown that for all cases the stability criterion depends upon a very simple rule: The *equilibrium point*, A_e, P_e, *is stable if the magnitude of the slope of Discount's buying strategy exceeds that of Crafty's selling strategy.* With the use of a simple model we have thus discovered a fact that might not otherwise be apparent: namely, the *slope or rate at which Crafty and Discount change price with respect to the quantity purchased* determines whether or not the equilibrium point is stable.

When the equilibrium point is stable, a slight disturbance brings about "forces," in this case psychological or economic forces stemming from Discount's and Crafty's business sense, which tends to restore equilibrium. On the other hand, for an unstable situation, a slight disturbance evokes forces that drive the price and amount even further away from equilibrium.

Final Comment

The example of this section is included not because of its importance in any actual analysis of economic systems, but rather because of three important points:

1. The analysis illustrates the phenomena of stability and instability. A system is operating normally (at an equilibrium condition). A slight disturbance or change in operation occurs. The system thereafter may
   a) Return to the equilibrium condition.
   b) Oscillate indefinitely.

c) Diverge further and further from equilibrium.

The first system is stable, the last is unstable, and the second is a situation on the borderline between stability and instability.*

* Whether we call this special case (Fig. 8–17) stable or unstable is arbitrary. If the oscillating system is useful, we often call it stable; if useless, unstable.

2. The example demonstrates the insight which we often can obtain by relatively simple analysis. Here we learned that stability depends only on the magnitudes of the slopes of the two strategies. A seemingly minor change in one strategy may convert a stable system to an unstable situation which results in disaster.
3. Finally, consideration of the example indicates what measurements we need to make on the system to determine stability—what features of the model are relevant. In our case, stability is determined by the two strategies. To simplify analysis, we have assumed these are represented by straight lines; actually the analysis is not really any more difficult if we assume curves.

Thus, the example is included because of the insight it yields into the stability phenomenon, rather than for a standard procedure that will predict the marketplace happenings to within a few percent. Our model omits all probabilistic considerations. It does not consider that Crafty and Discount will probably quickly guess each other's strategy and revise their original strategies accordingly.

Nevertheless, situations such as those pictured in our graphs are observed to occur in a free market economy. The interested reader is referred to *Economics*, 8th edition, by P. A. Samuelson, for a more extended discussion of the so-called "dynamic cobweb" of Figs. 8–17 and 8–18 and its relation, for example, to the statistics of the production of corn and hogs and the observed "corn-hog cycle."

8 | INSTABILITY IN PHYSICAL SYSTEMS

The detailed, quantitative study of instability has its roots in the exploitation of physical systems for man's benefit, for example, systems such as bridges, buildings, and electrical controls. The wheel with an off-center weight is a simple example. On level ground, two equilibrium conditions are obviously possible, as shown in Fig. 8–20.

Fig. 8–20. Stable and unstable weighted wheels.

From experience, we know that the equilibrium condition shown in Fig. 8–20a is stable and that shown in part (b) of the figure is unstable. A slight disturbance of the wheel when it is in the condition shown in (a) merely causes the wheel to rock back and forth until it settles into its original position. On the other hand, a slight disturbance of the wheel when the wheel is in condition (b) causes the wheel to turn further from its original posi-

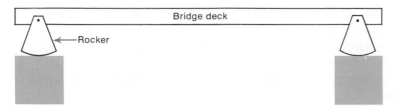
Bridge deck

←Rocker

*Fig. 8–21.
Bridge deck
on rockers to allow for
thermal expansion.*

tion so that the weight moves to the bottom. The wheel then rocks back and forth until it comes to rest as in condition (a). Thus, state (a) represents stable equilibrium, while (b) depicts an unstable situation.

A similar condition exists in a less obvious form in the following situation. In bridge construction, the road deck must be permitted to expand and contract as the temperature changes. To permit this to occur safely, the bridge deck design allows the bridge to vary its length at the points of support. In one case, the bridge deck was supported by rockers placed under the two ends, as shown in Fig. 8–21. While this bridge was in the process of construction, collapse occurred, which killed several workmen and incurred costly damages.

Collapse could have been avoided if the designer had been familiar with the principle illustrated in Fig. 8–20. The bridge deck is supported by rockers which can tilt as the length of the deck changes. This is shown in Fig. 8–21 and in detail in Fig. 8–22. The bottom of the wedge-shaped rocker is shown as a portion of a circular arc. The circle, from which the arc is cut, is centered at a point below the pin joint which connects the girders with the rocker. With no connection to the bridge deck, the only forces acting on the rocker are its own weight Mg and the reaction force from the earth (Fig. 8–23). The weight Mg acts from a point below the center of the circle, called the *center of gravity* of the rocker. The rocker itself is stable.

When the deck is pinned to the rocker, the much larger deck weight, W, acts *above* the center of the circle, to produce a situation similar to that in Fig. 8–20b, where the weight of the wheel is negligible in comparison to the applied weight W. It was the result of a condition of this type that a small disturbance sent thousands of dollars of material and effort crashing to the ground.

The effect of a disturbance on the equilibrium of the structure which has just been described can be examined in detail. For simplicity, we assume that the weight of the deck, W, can be considered as concentrated at a single point (the center of gravity). In Fig. 8–24, (a) shows a slight disturbance of the wheel when the center of gravity is *below* the center of wheel, (b) when it is *at* the center of the wheel, and (c) when it is *above* the center of the wheel. In all cases there will be a reaction force W' from the earth on the wheel. In both (a) and (c) the forces W and W' create a torque

Pin joint

Full circle

Center of circle

Portion of circle

*Fig. 8–22.
Details of
rocker and pin
joint.*

Pin joint

Center of bottom arc

Center of gravity

Mg

Mg

*Fig. 8–23.
Forces on
the unloaded
rockers.*

which tends to turn the wheel. In (a) the unbalance tends to rotate the wheel counterclockwise to *restore* it to a position where the center of gravity is at its lowest position. On the other hand, in (c) the unbalance tends to rotate the weight clockwise to turn the wheel further from its original position. In the borderline or *critical* case (b), a disturbance creates no torque.

This simple example, which can be readily demonstrated with rockers made of wood and a wooden deck loaded with weights, displays the same characteristics that we have met in our epidemiology and economic examples. In the stable configuration (a), a disturbance away from the equilibrium state evokes torques that tend to restore the system to equilibrium. In the unstable state (c), the evoked torques tend to drive the system further from the equilibrium state.

As mass is added to the bridge deck, both the magnitude of the effective weight and the height of the center of gravity (Fig. 8–24) are increased. With a sufficient added mass, the vertical position of the center of gravity is *above* the center of the wheel. At this point, the system passes suddenly and abruptly from a stable to an unstable condition.

An example of instability in a physical system which is of great engineering importance is aerodynamic or wind-induced instability of structures. This sort of instability has come to the fore only relatively recently as the result of sophisticated structural design of bridges and airplanes.

This effect can be illustrated with a small electric fan and a spring supported half-round section (Fig. 8–25). When the airstream from the fan is directed at the flat side of the section, the section starts to oscillate due to the periodic shedding of eddies. If the frequency of oscillation is close to the natural frequency of the spring support system, the oscillation of the section grows with ever increasing amplitude until either the section strikes the frame or the springs collapse.

Aerodynamic instability has been the cause of bridge and airplane failures in which millions of dollars and many human lives

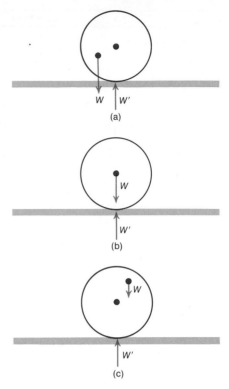

Fig. 8–24.
Slightly disturbed
wheel when the weight is below, at,
and above the wheel's center.

Fig. 8–25.
Sketch of model
showing aerodynamically
induced instability.

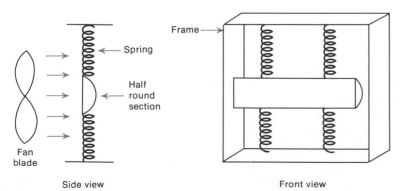

Side view   Front view

have been lost. For example, the collapse of the Tacoma Narrows Bridge was caused by resonance due to the periodic shedding of eddies. The bridge developed torsional oscillations with amplitudes up to 40° before it broke. The effects of such instability are particularly noticeable in the midwestern part of the United States. Here the broad plains allow steady winds to build up. The common result is highway stop signs snapped off by unstable oscillations, or the destruction of overhead telephone lines oscillating in the wind (called "galloping conductors" by engineers who attempt to introduce damping devices to control the amplitude of oscillations).

9 | USES OF INSTABILITY

Thus far we have discussed the phenomena of stability and instability, with examples in which instability represented an undesirable feature. We cannot, however, draw general conclusions about the overall desirability of any phenomenon. In some societies the social structure is often stratified. Individuals are not permitted to attain social positions which are considered "above their natural station in life." A society of this type may be considered highly stable.

In the colonies which preceded the establishment of the United States, a handful of people were confronted with the task of developing enormous natural resources. A growing population was beneficial. In terms of a population model, this growth was a phenomenon typical of instability. Under these conditions instability is a desirable feature.

As we have seen, a system which moves further from its initial state when it is disturbed is an unstable system. If this change is rapid, and produces a sharp demarcation between the initial and final condition, it may have useful applications. Perhaps the simplest application of this behavior is displayed by a mechanical switch, such as a light switch, shown in Fig. 8–26. In part (a), the switch bar is in an unstable equilibrium. It is impossible to keep the switch bar in this position. Instead, the switch bar prefers one

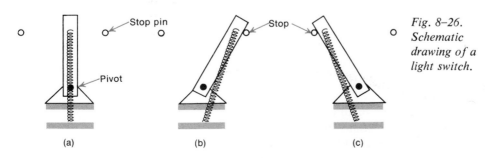

(a)   (b)   (c)

Fig. 8–26. Schematic drawing of a light switch.

of the two equilibrium positions on either side, shown in (b) and (c). By using the property of an unstable equilibrium, we produce a device that is definitely either on or off.

Thus, in a particular system we may seek either stability or instability, although in the latter case it is usually desirable to realize instability in a controllable fashion. For example, corporate management customarily attempts to realize continually increasing profits rather than to operate at a stable equilibrium point.

An even more obvious system in which stability is undesirable is the economic status of the world's population. Here it is obviously essential to move away from our present equilibrium point if significant amelioration of current poverty conditions is to be effected. Experience in many countries has shown, however, that the system is relatively stable; we cannot change the equilibrium point markedly by building new, improved housing for the underprivileged. Indeed, when this has been attempted in some sections of the world where illiteracy is common, the new housing rapidly degenerates into a new slum area.* Job opportunities and education must force a change in the structure of the system so that the equilibrium point becomes unstable.

* For example, in one such large development in Latin America, 50% of the bathtubs had been ripped out and sold within six months of occupancy.

In spite of these examples, we most often attempt to design and control systems in such a way that they are stable. Today in the United States a high percentage of the federal tax and economic policies are designed to maintain stability in the value of the dollar and to realize a controlled growth in national economic activity. Perhaps the most dramatic recent example of financial instability occurred in Germany in 1923. Following the fall of the German empire in 1918, food shortages, foreign occupation, costs of rebuilding, and inability to manufacture resulted in severe drains in gold reserves. As more and more paper money was printed, the value of the mark declined until, in 1923, four trillion marks were equivalent to one dollar. This highly unstable inflation destroyed the savings and foundations of the middle class, bankrupted many companies, and created the atmosphere in which Hitler was able to achieve power following the economic difficulties of 1932 (when 6,500,000 workers were unemployed in parallel with the severe depression in the United States).

Many other examples can be cited of systems in which stability is of vital importance—both in man-made systems, such as aircraft, and in natural systems interacting with man-made elements, such as ecological systems.

10 | FINAL COMMENT

In creating systems to serve man's ends or in coping with the world as we find it, it is not enough to seek a desired operating state. We must also concern ourselves with the stability of such a state.

If the state is unstable, a small disturbance initiates a rapid shift from the desirable state. Forces—physical, psychological, economic, or biological—are evoked which tend to move the system away from the desired state. On the other hand, if the desired state is stable, the evoked force tends to restore the system to the desired state.

The phenomenon of stability surrounds us. Traditionally man has studied stability in connection with physical systems where the stability phenomenon could be examined in its simplest form. But as we have observed in such diverse examples as epidemiology and economics, stability plays an important role in much broader areas of human activity. It is only recently that man has learned to apply the quantitative techniques of modeling to study the phenomenon of stability in other than physical systems.

Questions for Study and Discussion

1. What does the phrase "superhighways beget superhighways" mean? Explain in detail.

2. In the urban traffic example given in the text there are examples of stability and instability. Describe these situations and indicate for each one what the original disturbance was, and also the forces which tended to drive the system toward stability or away from stability.

3. Give an example of an unstable system which is not given in the text. Point out in your explanation the disturbance which started the instability, and the forces which tend to drive the system further from the operating state.

4. Describe briefly how the computer is used to maintain stability on a controlled stretch of highway.

5. What is one fundamental limitation of an expressway traffic-control system?

6. List the forces which tend to drive an epidemic toward instability.

7. List the forces which tend to drive an epidemic toward stability.

8. Construct a block diagram for an epidemic in which the following situation prevails:

N = total population
i = number infected
s = number susceptible
r = number recovered and immune
M_o = number who leave the population during the epidemic
M_i = number of new people who move into the population during the epidemic
R_i = rate of infection = $0.02si$

9. The textbook uses the rate of infection as $0.01si$. For a particular disease, how would this number (0.01) be determined? How could the use of computers provide a more reliable number?

10. Would a casual examination of the curve in Fig. 8–7 for the first 3 days give an indication that an epidemic is imminent? Why? How about the first 5 days?

11. Experience with graphic models indicates that for the first 5 days the graph in Fig. 8–7 is exponential. If it were actually exponential, it would grow

steeper on the 6, 7, 8, 9, and 10th days rather than flattening out as it does.

What happens during an epidemic to cause that flattening?

12. If all people who were infected recovered and were immune 10 days after the infection, would the graph pictured in Fig. 7–8 look any different? What would the graph look like if the time base were extended to 20 days?

13. Explain (as you would to a person your age who is not taking this course) how to construct a block diagram to find the population of a region if the following situation prevails:

N_o = original population

P_i = number of people moving into the region each year per 100 population

P_o = number of people moving out of the region each year per 100 population

R_b = number of children born each year per 100 population

R_d = number of people who die each year per 100 population

N_i = total population at the end of any given year

14. Suppose you had determined f_i and f_r in the epidemic model by examining statistics for a smallpox epidemic in Asia. Could these values be used to predict the course of a smallpox epidemic in New York City? Give reasons for your answer.

15. Which is more effective, as far as the critical num-

ber of susceptibles s_{cr} is concerned, increasing the recovery-isolation factor f_r by 50%, or decreasing the infectious factor f_i by 50%?

16. When Henry Ford brought out the original Model T "Tin Lizzie," he set a price that was fantastically low, and at the same time established a wage scale in his factory that was much higher ($5 a day) than elsewhere in industry. This brought about instabilities in the automobile business in general. Describe one or two of these. Why did they not snowball indefinitely?

17. Water impounded behind a dam is in unstable equilibrium, in the sense that if the dam should break, the water would rush downhill and perhaps do great damage to structures or people in its path. How do engineers employ this instability of impounded water for useful purposes?

18. Compare the desired state, the disturbance which moves the system, and the forces which tend to return the system to the desired state in each of the following situations:

a) a smallpox epidemic

b) a gasoline price war

c) an urban riot

d) a college student strike

e) soil erosion

f) an international "situation" between two countries which border each other

g) depletion of minerals in soil by "one crop" plantings

Machines and Systems for Men

9

I | INTRODUCTION

Any subject which has a great effect on the everyday lives of people soon leads to popular jokes, cartoons, and stories. Automation is no exception. One of the classic stories goes as follows:

Passengers for the Los Angeles-to-New York flight were all seated, the airplane doors were closed, and the motors started. The soft music was interrupted by a recorded announcement.

"Ladies and gentlemen, welcome to the *first totally automated* passenger flight. We will be landing in New York in four hours. The entire flight will be automatic. There is no pilot aboard, but the plane is under the complete control of an on-board computer. There is absolutely no cause for alarm. This computer is backed up by a communication link to a ground computer, which will watch your flight at all times.

"We assure you that there is absolutely no reason to worry. Every part of the aircraft-control system has been thoroughly tested. The aircraft engines, autopilot, and navigation system are absolutely reliable. So, sit back, relax, and enjoy your flight since nothing can go wrong, can go wrong, can go wrong, can go wrong, can go wrong. . . ."

While the story is extreme, it does illustrate a primary danger in modern technology. All too often, we become so enamored of gadgets, we lose all perspective on what the devices are really for. Technology should be used to give man a more pleasant, fuller, and richer life. Technology should relieve man of tasks which are tedious or difficult. It should perform tasks man cannot accomplish. It should be used to substitute for man when such a substitution is logical.

The problem in our story is obvious. Man is uniquely able to adapt to unexpected events. We cannot foresee *every* possible happening and build the machine to take care of all these eventualities. We must have men in the system—intelligent men who can respond logically to the circumstances.

While our story is extreme, we can think of many instances where technology is over-used. We call this situation "technological overkill." It is analogous to using a 40-foot crane to pick up a pencil from the sidewalk. As we look at the wonders of our age of technology, can we seriously feel that the average housewife's life is better because of the electric can opener? Or is the labor saved more than offset by the annoyance of trying to have it repaired when trouble develops?

Not too many years ago, auto water temperature was indicated by a thermometer on the dash panel. On hot summer days, in heavy traffic, the driver could see the temperature rising. When the temperature was approaching a dangerous point, the driver could shift to neutral and race the motor to cool it. Many of today's cars have only the "idiot light." There is no indication of water temperature until the boiling point is reached. Then the light goes on—often too late for the driver to do anything. This "modern" system does not match the machine and the man. Once the man could observe the way the temperature was changing and take corrective action. In the modern system, these human capabilities are not used.

In this chapter, we will look in more detail at this important problem of matching machines and systems to men. What should we ask technology to do? In other words, what are the things man cannot do? Where does he need to look to machines for help? How can we fit machines and men together in such a way that we maximize the benefits of technology?

To begin to understand some of these problems, we look at different shortcomings of a human being. In each of the following sections, we focus on a different human limitation and then consider how technology can be used to solve the problem.

2 | MAN AS A CONTROLLER

Control systems are those in which we are trying to keep an output signal in a desired position. For example, a common control problem is steering a car. Here the driver observes the location of the car in the road and then makes corrections by turning the

Fig. 9–1. A simple resonant system.

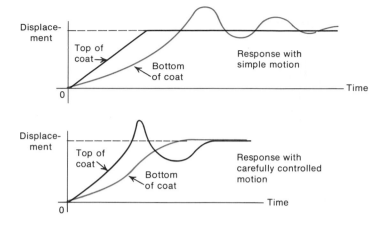

Fig. 9–2.
*Damping the
resonance.*

Fig. 9–3.
System analogous to Fig. 9–1.

steering wheel. As he makes turns, he tries to keep the output signal (the car position) in a desired location.

Figure 9–1 shows a man involved in a simple task. He is holding his jacket at the top. He wants to move the jacket sideways to a new position. If he merely moves the top to the new, desired location, the bottom oscillates around its final position. Eventually the oscillation dies down and the system comes to rest in the new position. We can plot the position of the bottom of the coat (Fig. 9–2).

If the man does this over and over again, he soon learns how to avoid the oscillations. He finds that he can give the top an extra push to get the bottom moving. Then he jerks the top back hard to brake the bottom. With a little practice, he can learn exactly how to move the top so that there are no oscillations at all (Fig. 9–2).

We find exactly the same problem in construction work (Fig. 9–3). When a huge crane moves a large weight, it may be very important that the load *not* swing back and forth. Such oscillations may damage nearby buildings. Unfortunately, the required motion of the top depends on the weight of the load. (This could have been shown by filling the jacket pockets with heavy objects.) In our crane example, the human operator can't practice a few times with each different load. Consequently, we have to use automatic technology in place of the man. *The control task is too difficult for the human being.*

Control of an Unstable System

Man also has trouble controlling an unstable system. In Fig. 9–4, the broom stands, upside down, on the man's hand. As the broom starts to fall, the man tries to move his hand horizontally in such a way that the broom stays up. After a few seconds, he loses control, and the broom falls. Here the system we are trying to control is unstable. In physics, we would call this an "inverted

Fig. 9–4.
*A man trying to
keep a broom upright.
The same experiment can be
done with any long stick
(e.g., a blackboard
pointer).*

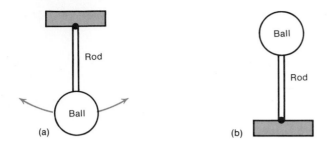

(a)

(b)

Fig. 9–5.
Pendulums
a) *Normal pendulum.*
After a slight movement
of the ball left or right, it
oscillates back and forth around
the vertical position. The ball finally
comes to rest at the starting position
(before the original disturbance).
b) *Inverted pendulum. As long as the*
rod is exactly vertical, the ball
will not move. The slightest
disturbance, however, results
in the ball rapidly falling.
This system is unstable.
It cannot recover from a
disturbance, no matter
how small.

pendulum" (Fig. 9–5). Without control, any small disturbance causes the ball to fall and it never returns to its original position.

If technology is substituted for man, we can control such a system without difficulty. Indeed, we can control several brooms simultaneously. Figure 9–6 shows a system which was actually built at Stanford University to demonstrate the control of such unstable systems. In this case, there are two inverted pendulums (rods A and B). The position and velocity of each rod are measured at all times. These signals are used to move the cart back and forth in such a way that the two rods are kept in an upright position.

The idea of controlling an unstable system is of practical importance. Airplanes have an unstable oscillation at a very low frequency and the altitude of the airplane tends to change at a very low rate (a cycle every ten seconds or so), with the amplitude of the oscillation increasing with time (Fig. 9–7). The oscillation in this case is so slow that the human pilot or the autopilot can easily correct for it and keep the altitude constant.

When the unstable frequency is higher, the human being is unable to respond fast enough to keep control. Figure 9–8 shows another example of an inherently unstable system. Here a satellite carrying men is tied by a rope to another space vehicle. (The latter might be the final stage which is separated from the main vehicle after the desired orbit is reached.) One of the problems associated with men in orbit for long periods of time is the effect of "weightlessness" (e.g., the calcium normally in the bones tends to dissolve). In order to create an artificial gravity, the two vehicles are spun as indicated in Fig. 9–8. This spinning motion causes an effect on the man which is like gravity. The speed of rotation is con-

Fig. 9–6.
Two rods on a movable cart.

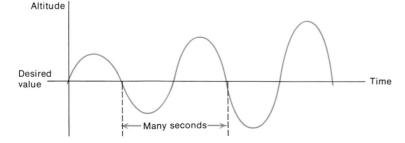

Fig. 9–7.
Unstable motion
of an airplane. Because
of the very low frequency
(the long period), the
oscillation is easily
counteracted by
the pilot.

trolled by reeling in or playing out the rope between the two vehicles. The control problem is to change the rope length in such a way that the rotational speed is constant.

Such a system is unstable. Furthermore, the frequency of oscillation is so high that a man cannot control the system. The rope has to be changed in length by a machine operated by a computer, which rapidly calculates the required rope length and then drives a motor which adjusts the rope.

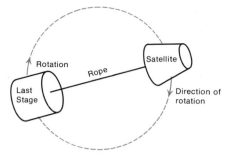

Fig. 9–8.
Two tethered satellites.

Control of a Hovering Aircraft

A VTOL (vertical take off and landing) aircraft can be positioned by changing the direction of the jet exhaust. When the aircraft is stationary over one spot (Fig. 9–9), the jet exhausts point straight down. The reaction force lifting the aircraft is just enough to offset gravity.

Fig. 9–9.
One vertical take-off
and landing aircraft has
a jet exhaust which can be
rotated. When the jets exhaust
downward and the motors are going
full, the aircraft rises straight up. Once
the desired altitude is reached, the jets
are rotated so they give a forward
thrust. The decrease in lift is
offset by the usual flow of
air around the wings.

Suppose the aircraft is hovering at one spot and the pilot wants to move horizontally to a new location (Fig. 9–10). Through his controls, he can apply a force tending to rotate the jet exhaust. This force eventually causes the exhaust tubes to start turning.

This system can be modeled as shown in Fig. 9–11. The pilot's signal, asking for lateral motion, causes this movement through

GROUND

ORIGINAL POSITION   DESIRED POSITION

Fig. 9–10.
Changing
the location of our
hovering VTOL
aircraft.

Fig. 9–11.
Model for the horizontal position of the VTOL aircraft.

four integrations. This model is derived directly from Newton's laws* in physics.

In Fig. 9–12, a man turns a dial to adjust the input signal, while he watches the meter measuring the output. The output is initially constant at -1. The man's job is to apply an input which will bring the output to $+1$ and hold it there. When we try this in the laboratory, we find that the man just can't learn to control the output. A typical set of trials is shown in Fig. 9–13. Apparently no matter what signal the man applies, the output runs away. It is not surprising that the VTOL aircraft is so difficult to fly.

When an automated system is used to control the output, the necessary signal to give the desired output is calculated from the equations describing the system. In this particular example, it turns out that the required input signal is as shown in Fig. 9–14. There just is no way the human being could learn to decide on this input signal. (Fortunately, there are some problems which we can solve with mathematics that we cannot solve by guesswork or by practicing on the system.)

The examples just mentioned are three cases in which man is not able to perform control tasks. In each situation, the tasks can be so difficult that man can never succeed, no matter how long he practices or what special training he receives. Part of the limitations of man come from the behavior of man as a system. As we saw in Chapter 1 and again in 6, the human controller has a dynamic response. For example, there is always a time delay of 0.2 second or more. Man's ability to predict a future signal is limited.

These characteristics can be traced to the time for signals to go from the eye to the brain and then to the muscles, and also to the type of computations made in the brain. As a result, man is unable to control a system which is moving too rapidly. If the oscillation, for instance, is at a frequency greater than one cycle/second, man tires rapidly. Above 1.5 cycles/second, he loses control completely. Even when the signal is changing slowly, man may be unable to control the system if the necessary control system is very complicated. The VTOL aircraft example illustrates this limitation. In the 1950's, when power steering for cars first ap-

*
Isaac Newton (1642–1727) was a remarkable man. After being a good but undistinguished student in college, he graduated at the age of 23. He wanted to stay on, but Cambridge University was closed for two years because of the plague. So he spent the two years discovering new theorems in math (and numerical analysis), developing the calculus (the most important math development in many centuries), discovering the law of gravitation, explaining the motions in the solar system, developing his laws of motion, inventing the reflecting telescope, discovering that white light can be broken down into all colors, and so forth. After the University reopened, Newton was made a professor at the age of 26. He lived almost 60 more years, but never again approached his early productivity.

We might draw two conclusions: (1) Schools should close more often. (2) Don't waste creative men by making them professors.

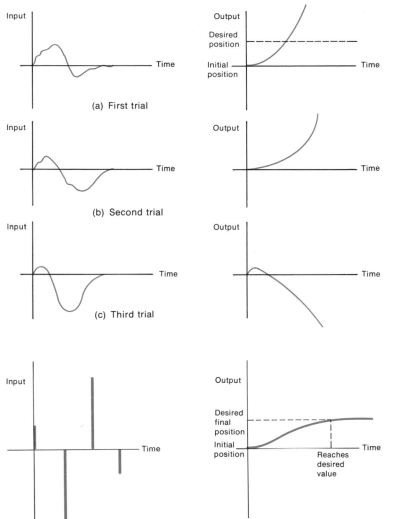

Fig. 9–13.
*Typical results
when a man tries
to control the system
of Fig. 9–12. In the first
trial, he applies a positive
signal, then a negative one to try
to brake. The output runs away in a
positive direction.*

*In the second trial, he uses a smaller
positive signal and more braking. The
system still runs away positively.*

*In the third trial, he uses a very little
positive signal and a lot of braking—so
much braking that the output now runs
away in a negative direction.*

*No amount of practice seems to
lead to the right combination of
positive signal to get the output
started and negative
signal to brake.*

Fig. 9–14.
*An input signal
which does give the
output we want. A positive
pulse is used to start motion. A
big negative pulse brakes. A positive
pulse is used to offset the braking in part.
The input signal ends up with a small
negative pulse.*

*This kind of input signal
is not at all obvious. No pilot
is likely to learn to fly this
way. An autopilot is
absolutely essential.*

peared, some of the steering systems for trucks became unstable when the truck made a sharp turn at moderate speed. The system was so complex, the driver was unable to bring the vehicle under control. Several severe accidents resulted before the power steering system was changed.

Thus, man may be unable to control a system because of the speed at which signals are changing or because of the complexity of the control system.

3 | MAN IN COMMUNICATION

The difficulty of accurate communication is illustrated if we have a small group of people, each speaking two or more languages. We give the first person a sentence in English. He relays it to the second person in a different language (say French). After a small number of translations, we come back to English. Very often there is no obvious relation between the sentences at the beginning and end.

One problem is that each person interprets the sentence in his own way. As the sentence travels from person to person, even the main idea may change radically. The chance for change is increased even further by the fact that a listener may not hear a word or phrase correctly. We are particularly likely to misunderstand a word if we expect a certain message. Finally, errors arise because other noises or sounds interfere with our hearing correctly. The unwanted sounds "mask" or drown out the message.

In ordinary conversation between two people, we use enormous amounts of redundancy to correct such errors. For example, we repeat the same thought several different ways if it is important. We use inflections in our voice, facial expressions, and even hand motions to help communicate. These actions enrich the conversation, and they also help to correct misunderstood words or phrases.

In spite of these redundancies, voice communication between people is filled with errors. How often have we been told a telephone number, address, or other information and found later we had the wrong data. In some cases, we were told wrongly; in others, we failed to remember; but often we just misunderstood.

In today's complex systems, we often cannot afford any chance of error. Stock prices given to a broker must be correct. Otherwise, he will be buying or selling foolishly. A few such mistakes and our broker is bankrupt (or rich). As another example, when we are controlling high-speed trains in a railroad network, we cannot allow errors as we send information on the position and speed of each train to a central computer.

Can we communicate data with no chance of error?

Automatic Error Correction

When we want reliable communication, we ordinarily send information in binary form. That is, we use a sequence of 0's or 1's. For example, a typical signal might be:

$$1\ 0\ 1\ 1\ 0\ 1\ 0\ 1\ 1\ 1\ 0\ 1\ 0\ 1\ 0\ 0$$

This sequence of 1's and 0's, reading from left to right, carries the information of the signal. Electrically, this signal can be sent as shown in Fig. 9–15. Regularly timed pulses are sent. During

Fig. 9–15. Pulse train representing the binary signal.

Pulse intervals

Fig. 9–16.
Received signal when
noise is added. (Only the first
six pulse intervals of Fig.
9–15 are shown.)

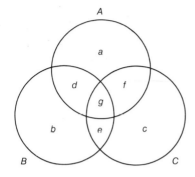

Fig. 9–17.
Three overlapping circles
with the regions a through g defined.

any space for a pulse, a pulse being present represents a 1, no pulse signifies a 0.

When noise is added during transmission, the received signal has the appearance of Fig. 9–16. During the second pulse interval, the signal should be zero, but the noise happens to be positive during this period. There is the possibility that this looks like a pulse to the receiver. If these first six binary digits

$$1\ 0\ 1\ 1\ 0\ 1$$

represent the number 45 in the message, the error in the reception of the second digit means the receiver interprets the number as

$$1\ 1\ 1\ 1\ 0\ 1$$

or number 61 in the decimal system. (We are assuming that the above are binary numbers, converted in the usual way to their decimal equivalents.)

In order to avoid this sort of occasional error, we can introduce redundancy as follows:

1. The signal to be transmitted is first grouped in sets of four digits. We initially consider the first four

$$1\ 0\ 1\ 1$$

2. Next we draw a set of three overlapping circles called A, B, and C (Fig. 9–17). In this figure, there are seven distinct, closed regions. Two such regions are g common to all three circles and f common to A and C.

3. In place of $d\ e\ f\ g$ we insert *in order* the four numbers of our message, 1 0 1 1 (Fig. 9–18).

4. We now fill in the numbers for a, b, and c by choosing each as either 0 or 1 in such a way that the total number of 1's in each circle is even. In other words, in Fig. 9–18, circle A already includes three 1's; hence, we must choose $a = 1$ to make the total number of 1's even (here four). Similarly, $b = 0$ and $c = 0$ (Fig. 9–19).

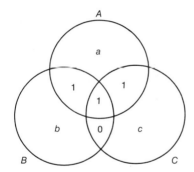

Fig. 9–18.
The digits of the signal are inserted.

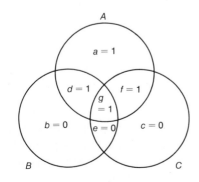

Fig. 9–19.
Determination of transmitted signal.

5. We now transmit the seven-digit sequence,

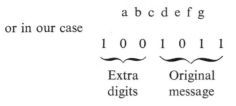

or in our case

$$
\underbrace{1\ 0\ 0}_{\substack{\text{Extra} \\ \text{digits}}}\ \underbrace{1\ 0\ 1\ 1}_{\substack{\text{Original} \\ \text{message}}}
$$

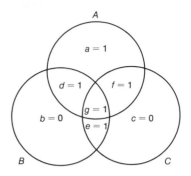

Fig. 9–20. Received signal including error in e.

Each block of four digits in the original message is handled in this way. In other words, for every four message digits, we actually transmit seven digits, three more than necessary.

These three redundant or extra digits permit *automatic correction* at the receiver of any single error in the seven-digit block. The receiver must, of course, be familiar with the *coding* used at the transmitter. Then at the receiver we reconstruct the three-circle diagram of Fig. 9–19. If any one digit is received incorrectly, one or more of the circles contains an odd number of 1's.

As a specific example, we can return to our earlier problem and assume the second digit of the message is wrong. That is, the receiver obtains

$$
\underbrace{1\ 0\ 0}_{\substack{\text{Redundant} \\ \text{digits}}}\ \underbrace{1\ 1\ 1\ 1}_{\substack{\text{Four digits} \\ \text{of message}}}
$$

The receiver checks the validity of this message by constructing the diagram of Fig. 9–20.* Inspection reveals that there is an odd number of 1's in circle *B* and also in circle *C*. Therefore, if there is only one error, it must be the element common to circles *B* and *C*, or element *e*. Hence, the received signal can be automatically corrected to

$$
1\ 0\ 0\ 1\ 0\ 1\ 1
$$

The communication code described above (known as the Hamming code) uses three extra digits for each four message digits. The same procedure can be used if we draw four regions overlapping in all possible combinations (this is a complicated geometric construction). There are then four extra digits for each 11 message digits. Similarly, five extra digits compensate for 26 message digits. As the message block is made longer and longer, however, there is a growing possibility of two simultaneous errors: a situation which the system cannot correct.

This example is interesting because it is used in practical communication systems. It is primarily of value, however, to indicate the simple way in which technology uses redundancy to improve communication among men and machines. With technology, we can have better performance than is possible in direct man-to-man communication. With the code described above, we can *auto-*

---

*Of course, electronic equipment does not construct a diagram and look at it. Instead, the receiver is designed to measure $a + d + f + g$, $b + d + e + g$, and $c + e + f + g$. Each of these three sums should be even. In the case of our received signal, the three binary sums are 1 0 0 , 0 1 1, and 0 1 1; the last two are odd; the error is in the element common to *B* and *C*, but not to *A*, or element *e*.

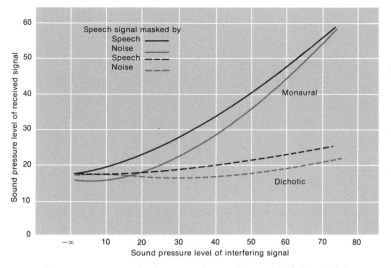

Fig. 9–21.
*Understanding
speech in the presence
of noise. The vertical axis
shows the intensity of speech
which can just be understood when
noise is present. The horizontal
axis shows the intensity of the noise or
a disturbance signal.*

*The monaural curves refer to the test
with one ear only used. Two cases are
shown: one when the disturbing signal is
speech, the other when it is a random
background noise. (From Man-Machine
Engineering, by Alphonse Chapanis.
Copyright 1965 by Wadsworth
Publishing Company, Inc.
Reprinted by permission of
the publisher, Brooks/Cole
Publishing Company,
Monterey,
California.)*

matically correct any single error in each set of digits. With more complex codes, we can correct any one or two errors. Thus, we can build communication systems which are essentially error-free.

Communicating in Noise

In parallel with his use of redundancy, man has a remarkable ability to "hear" only particular portions of the sound reaching his ear. For example, two men can carry on a conversation even in an extremely noisy room. Figure 9–21 shows the human capability. We can understand even when the noise intensity is very much larger than the message sound. This characteristic is called the "cocktail party effect."

In spite of these remarkable capabilities of the human being to communicate in the presence of interfering signals, if the noise becomes too large and the signal too weak, errors occur. If communication is to be accomplished under such circumstances, it is necessary to slow down the rate and limit the vocabulary being used. One procedure which has proven very effective is called *word spelling*. Each word in the message is spelled, but with each letter represented by an agreed upon code word. The International Civil Aviation Organization has developed the following code for such purposes:

Alpha	Hotel	Oscar	Victor
Bravo	India	Papa	Whiskey
Charlie	Juliet	Quebec	Xray
Delta	Kilo	Romeo	Yankee
Echo	Lima	Sierra	Zulu
Foxtrot	Mike	Tango	
Golf	November	Uniform	

To send the word STOP, we say "Sierra Tango Oscar Papa."*

*
In the choice of the words in such a code, we desire to meet several different criteria. Obviously each word must be readily identified with its letter (we would not use *Cease* for C both because the initial C sounds like an S and because of the ease of confusion with Sierra). Furthermore, the words must be easily recognized when spoken by people of many nationalities. Once the code is agreed upon, however, communication is made easier because the listener has to decide what the received word is from among only 26 possibilities.

Fig. 9–22.
*Signal detection
problem.*

Fig. 9–23.
*Processing
of received
signals.*

Even with these techniques, there are cases in which man's ability to detect signals in noise must be helped by technology. For example, when radar measurements are made of the planet Venus, a short burst of energy is radiated from an antenna pointing at the distant planet. This radio energy travels to Venus, a small fraction of the energy hits that planet, a small fraction of that is reflected back toward earth, and yet another small fraction of that reaches the antenna. Because of the noise signals which are also received by the antenna, the noise may be a thousand times larger than the signal.

We are then faced with the problem shown in Fig. 9–22. On an oscilloscope, the received signal simply looks like noise. Somewhere buried within this signal, however, there is a very small component which, if it were magnified, would look like part *b* of the figure. This is the echo.

If a regular series of transmitted pulses are used, each received signal (r_1 and r_2) has the form shown in Fig. 9–23. Comparison of two different received signals reveals that the noise in one is unrelated to the noise in the other; however, the echoes are the same in the two cases. Thus, if we multiply the two received signals together ($r_1 r_2$) at similar instants of time (Fig. 9–23), the noise tends to average out (if r_1 is positive, r_2 is as likely to be positive as negative). The echo parts of the two signals are exactly alike, however: when the echo signal in r_1 is positive, so is that part of r_2. When this multiplication process is carried out with signals received after many different transmitted pulses, the echo part of the product becomes larger and larger compared to the noise part. With enough manipulation of the data in this fashion, we can identify echo pulses which are very much smaller than the average noise level.*

Thus, technology can be used to detect a signal of known shape which is swamped by a very large noise signal. This use of electronic equipment has been developed only during the last twenty years, primarily as a result of attempts to build radar equipment to obtain echoes from distant planets and very small objects in the sky.

In communication man works under very severe limitations Errors do occur. Noise tends to mask the signal. In addition, as we mentioned in Chapter 6, the rate at which man can communicate is limited. If we wanted to send the latest prices of 1000 stocks to a friend across the country, we certainly would not call him on the phone and read them one-by-one. Even at four a minute (with time allowed to repeat in order to avoid errors), the job would take over four hours. By that time, the prices would have changed. We would use automatic equipment which could easily send the information in a few minutes.

* This problem of finding a signal in noise also occurs in other situations. When an attempt is made to determine whether a person is dead (for example, in order to remove a kidney or the heart for a transplant operation), one of the criteria is that the electroencephalogram should show no brain waves, even within the normal noise of the instrument. In a different direction, current attempts to detect signals from possible civilizations in other parts of the universe require the detection of signals of unknown frequency amidst the noise of space and the receiver. Economists and businessmen frequently search for periodic or sinusoidal variations in profits or economic indices in order to detect oscillations which should be taken into account in business decisions.

4 | MAN LIMITED BY ENVIRONMENTAL NEEDS

Man is seriously limited by the environment he needs. This limitation has been extensively studied in the manned space program. The decision by the United States to focus efforts on manned exploration of the moon meant that the space vehicle had to provide an environment in which man could survive and work for a period up to two weeks. As a result of the need for the presence of man, the interior of the space vehicle was supplied with oxygen and nitrogen so that men could breathe without helmets. Temperature within the vehicle was controlled to avoid extreme cold or heat. In a longer space mission, artificial gravity will be used to avoid weightlessness. In other words, the presence of man leads immediately to a much more complex space vehicle. In addition, of course, we then place great emphasis on returning the vehicle safely to the recovery area.

These same environmental needs of man are severe limitations in other places. Robots have been developed to work in radio-

*Fig. 9–24.
How stressful
is long confinement
in small places? Ask an
astronaut or any commuter.
But what is the very least amount
of room needed? To find out,
Dr. Reginald Whitney of England's
Medical Research Council has been
putting volunteers in cages like this for
24 hours at a time. The man's position
is varied between lying and sitting while
he is given tasks to do and his mental
alertness is checked. Meanwhile, heart
rate and skin temperature are monitored,
urine is analyzed, and psychological
tests are administered. A subject can
kick his way out of the cage.
None did, which augurs well
for crowded space capsules
and led experimenters to
suggest that minicars
could be made more
mini.* (Science and
Technology,
Jan. 1968)

active areas—for example, to position objects within an operating nuclear reactor. Robots are currently being made to carry fire hoses and chemicals into burning buildings where there is a danger of the structure collapsing. When such fire fighting equipment is in general use, we should avoid some of the tragedies besetting firemen today.

Part of the "environment" for man is his acceleration. (Gravity is one example.) Here the need for human comfort limits the rate at which trains can be accelerated to a high speed or elevators can descend. If the acceleration is too great, man is uncomfortable. Repeated acceleration and deceleration (as in a subway train reaching high speed between stops) can cause severe discomfort and eventually nausea. (Usually the limit in a subway or bus arises first from the desire not to injure passengers who are standing, or trying to stand.)

This need for a pleasant environment also limits high-speed trains riding above ground. Most people like windows in a train. If there are windows and the train moves at 300 miles per hour, there must be an appreciable amount of clear space on either side of the train. If trees, houses, and so forth are too close to the windows, they will move by so fast the passenger will become carsick. (One designer proposed to use artificial windows, through which movies show a landscape moving by at a reasonable speed, perhaps 60 mph.)

Thus, the environment for man includes all of the signals which he senses, some of which are noise, temperature, pressure, acceleration, speed. We also must include the air he breathes and the stress under which he works. Man is remarkably adaptive to changes in this environment. He can work, for example, in environments with a very wide range in temperatures. Man can even perform useful work during high accelerations in rocket travel. When the limits are reached, however, machines must be used to replace men.

5 | MAN AND SENSING

Man's personal tie to reality is through his senses. Human observers can sense, or perceive, many quantities. Light and sound are particularly significant since they provide us our best means of personal communication, by sight and hearing. The human body has a large number of very refined sensing devices. They are built into some two square yards of body surface which make up the interface between our internal and external worlds. Specialized sensors that respond to different signals are scattered over this surface. It is through these sensors, including the eyes and ears, that all information about the external world comes to consciousness.

In addition to the eyes and ears, there are other sensors distributed over the body, such as those for the detection of pressure, heat, and cold. In addition, there are internal sensors which inform us of the positions of various parts of the body (for example, we know where our arms are even with our eyes closed). There are sensors which tell us how fast a limb is moving and the difference between up and down. Many people have a very good sense of compass direction which helps them, for example, to estimate how much they have turned while blindfolded. Some people have a remarkable sense of elapsed time. The list goes on and on, totaling many more than the five senses we are customarily said to have.

Many of these senses are extremely delicate or sensitive. Our normal sense of hearing can detect a sound so feeble that it moves the most sensitive part of the inner ear only some 10^{-11} centimeters. This distance is roughly one-thousandth the diameter of a hydrogen atom! If this sheet of paper is four-thousandths of an inch thick, then it is a billion times thicker than the distance the inner ear mechanism moves in responding to a faint sound.

Signals Man Cannot Sense

There are many quantities, however, which man cannot sense directly. Radio waves and X-rays pass through our bodies without any sensible effect. Try as we may, we cannot sense directly the television signal sent out from the broadcast station or the high-

306

Fig. 9–25.
View of the Bucks Lake
area of the Sierra Nevada
was obtained with panchromatic
film, meaning film that is sensitive
to the entire visible spectrum.
Such a photograph is particularly
useful for estimating the
density of vegetation
and for identifying
certain species of
vegetation. (Dr.
R. N. Colwell,
School of Forestry,
University of
California at
Berkeley)

pitched screams which certain bats use for locating food. We can and do, of course, create sensing devices that transform an insensible signal into one that can be sensed. Television sets, for example, change broadcast signals into light and sound.

Some animals can sense signals which man cannot. Porpoises and some fish can hear sounds too high in pitch for man to hear. Some migratory birds and animals, as well as homing pigeons, are believed to sense the earth's magnetic field directly—a useful ability when navigating to places beyond the line-of-sight.

Among the most impressive modern sensors are those used in space vehicles or high-flying aircraft to observe the earth. Photographs, taken with special cameras, show a surprising amount of detail, as Fig. 9–25 indicates. The ability to inspect equipment, airfields, missile installations, and the like from space is, according to government experts, one of the major hopes for effective disarmament. Countries will seldom, if ever, allow on-site inspections by observer teams even to confirm that treaty terms are being kept. At the same time, nations are reluctant to sign agreements without the ability to confirm that all parties are conforming. Overhead photography from satellites and, in some cases, high-flying aircraft provides a technique for inspection which is often acceptable to all parties. The ability to know that no other country is deploying its missiles, ships, and other forces to attack can indeed be an influence for peace in the world. Satellite photography is useful too in weather forecasting, surveying of natural resources, and in locating sources of pollution.

Selective Sensors

Sensors in space and in high-flying aircraft are able to reveal much more than the usual camera might show. For example, objects of a specific color can be picked out and highlighted in a picture by using a filter of that color in front of the camera lens. The filter allows light of the desired color to pass through, but absorbs light of other colors. In a photographic print, the object can be made to appear light against a dark background.

A series of nine pictures of the same scene taken at the same time with a nine-barrel camera is shown in Fig. 9–26. Each barrel had a different filter so nine different views result. (The photos are

Fig. 9–26.
Multiband exposure
was made with a multiband
camera. The wavelengths of energy
(that is, colors in the visible part)
represented range from .38 micron
(top left) to .9 micron (bottom
right), which covers not only
the visible spectrum, but
also part of the ultraviolet
and near-infrared regions.
(Dr. R. N. Colwell,
School of Forestry,
University of
California at
Berkeley)

Fig. 9–27.
The same picture
taken with two different
filters. (Dr. R. N. Colwell,
School of Forestry,
University of
California at
Berkeley)

printed in black and white even though they were taken with colored filters.) Note that the various trees, buildings, and ground features show up quite differently in the different photographs. This happens because each object reflects different colors of light. In fact, objects tend each to have a *signature:* a pattern of light that they reflect. This can be determined by shining white light (which contains all colors equally) on the object and then measuring the color or colors of the light reflected from the object.

The importance of the "tone signatures" of objects is very well illustrated in Fig. 9–27, which shows two photographs taken with different filters of the same scene. The left photograph shows underwater detail to about a depth of 20 feet and highlights a road through the upper part of the picture. The right photograph emphasizes waterways rather than the road and shows certain plants and vegetation as bright white areas. The combination of pictures gives much more information than either one alone.

It turns out that objects reflect not only light, but heat and radio waves as well. In fact, light, heat, and radio waves are all closely related; they are all examples of *electromagnetic* waves. The only difference among them is their frequency.

Thus, objects reflect infrared (heat) and radio waves just as they reflect light, that is, some frequencies are reflected and some absorbed. Objects have an infrared and radio signature just as they have a visible light signature. With proper sensors, infrared and radio pictures can be obtained. These can reveal more than visible light alone can.

Infrared photography makes use of special film sensitive to the infrared portion of the spectrum. One principal advantage of infrared photography is that these waves penetrate smoke and haze and so can be used under conditions where ordinary photography would give poor results. Another use is in studying land features and vegetation. Infrared pictures taken at night will show objects which are hotter than their surroundings. In Fig. 9–28, three infrared pictures of the Yosemite Valley are shown. In one

Fig. 9–28. Infrared pictures of part of the Yosemite Valley. (Dr. R. N. Colwell, School of Forestry, University of California at Berkeley)

the camera was equipped to show fires in the valley. Most are campfires, but even single charcoal briquettes were visible. The second picture shows also the heat from certain vegetation, recorded at lower frequencies. A daytime infrared picture of the same site is also shown in Fig. 9–28. Timber shows clearly here by reflection of the sun's infrared radiation.

All of the above demonstrate that appropriate sensors can reveal a great deal about the real world; they can extend man's own senses, enabling him to perceive objects and events beyond his native capacities. This possibility opens many avenues. One example is food-resource management to make adequate food available. Today only about 10% of the world's land area is cultivated; an additional 21% can be farmed. To make food needs and food production balance, more information is required in order to determine what crops to grow where and what the likely yields will be. The techniques discussed above, adapted for surveying, can help provide this information. They can provide, for example, mapping of soil and water temperatures, mapping of surface water, mapping of disease and insect invasions, assessment of crop vigor, determination of soil type and soil moisture, and mapping of major soil boundaries. This information could be invaluable in locating and planning usage of land for food production. More broadly, the information could provide data quickly on where existing crops are threatened, what crop resources are actually available in famine areas, and on what crops could be grown where. Sufficient manpower is not available to collect such information by personal inspection.

Final Comment

Aerial and space photographs are just one example of the use of technology to extend man's sensing ability. We can give other illustrations. For example, we often need to measure temperature with an error much less than one degree. Man can certainly tell when an object is hot or cold; he cannot tell whether a temperature is 121.2° or 121.3°. Yet this accuracy is necessary if the Apollo navigation system is to work properly.

Automated factories require that we sense the condition of products being used in the manufacturing process in order to be sure of the quality of the output product. Inside a boiler of an electric generating station, we must measure temperatures continually to avoid explosions. In a high-speed train system, tragic collisions are prevented by sensing the distance between trains. In the automated highway of the future for intercity travel, cars will be moved at high speed to their destination. Light beams (from lasers) will be sent out frontward from a car to measure the distance to the next car ahead. We must detect at once a decrease in this distance in order to brake soon enough to avoid accidents.

Finally, air-pollution sensors throughout the city warn of danger early enough to shut down incinerators, ban auto traffic, or take other safety action.

Thus, whenever we build a technological system to replace man in a particular task, that system must include sensing. In sensing, we measure the signals which represent the important behavior of the world with which the system interacts. Sensing is an important area in which devices are often much better than man.

6 | PROSTHETICS

In the preceding sections of this chapter, we have seen a few ways that technology is used to extend man's abilities: to do things which man cannot do. In a parallel direction, we try to use technology to give a normal life to handicapped people. One measure of a civilization is the compassion that society has for its less fortunate members. Among this group are those who have lost arms or legs, are deaf or blind, or are in some way physically unable to pursue normal activities.

In recent years, widespread publicity has been given to the attempts to develop an artificial heart and an artificial kidney. The transplant of a human heart emphasized the pressing need for an artificial heart, because the possibility of finding the more than 100,000 heart donors required annually in the United States alone is obviously remote.*

The high cost and scarcity of artificial kidney machines have accelerated engineering attempts to develop a portable, low-cost device that could be used in the patient's home. More than 20,000 people die needlessly each year in the United States because kidney machines are not available. These could be provided at a national cost of about 100 million dollars. The problem here is the fascinating one of national priorities. Should we spend 100 million dollars more each year? If we decide to spend 100 million dollars more on health each year, should we put this money into kidney machines? Who is to decide the answers to these questions? Shall we leave the decision to Congress? How do we avoid simply giving in to the group which makes the most noise? Such questions are of such fundamental importance for the future of our society that the proposal has been made that everyone's home phone should be tied to a computer network so that the people could vote on questions of national priorities.

From such broad social questions, let us return to our consideration of *prosthetic* devices: devices to replace parts of the human body. Perhaps the most familiar are the technological devices to help people who have impaired senses, particularly hearing and sight.

*

Even if enough donors could be found so that all weak hearts could be replaced by transplanted human hearts from individuals who died of other causes, there would be an argument for the use of an artificial heart instead. The human body tends to reject organs transplanted from other human beings more than artificial mechanical devices. The principal argument for using transplanted hearts is that engineers are not at all sure they can build an artificial heart which will work reliably for perhaps 15 years.

Fig. 9–29. Operation of a hearing aid.

Fig. 9–30. Diagram of ear parts (*cochlea is shown straightened out; it is actually a three-turn spiral about 35 millimeters long*).

Hearing aids are exceedingly common among older people. The hearing aid consists simply of a microphone to detect the sound waves and an electronic amplifier and small loudspeaker to generate a sound signal which is an enlarged replica of the incoming signal (Fig. 9–29). The system is powered by batteries which are often concealed inside the wearer's clothes or behind the ear. A volume control permits the man to turn down the sound when he is anxious for quiet (as during a boring lecture or conversation) or when he is in a particularly noisy environment.

The successful use of a hearing aid requires that the man have at least a little hearing capability left: he cannot be totally deaf. All the hearing aid does is amplify the sound signals so that they can be heard by the listener. The man still detects the sound signals in the usual way.

The human ear is a complex device which operates in the following general manner. Sounds are really pressure changes in the air. When the sound signal reaches the eardrum (a thin membrane), the pressure changes cause the eardrum to vibrate or move back and forth: the louder the sound, the greater the back and forth motion; the higher the pitch, the faster the motion (Fig. 9–30).

Just behind the eardrum are located various bones formed to act as a lever. This lever multiplies the pressure changes and reduces the amplitude of motion by a factor of about 16 (just as a lever permits a man to lift large weights, or two people of radically different weights to balance on a seesaw). The stronger pressure changes are then sent on to the liquid filled inner ear.

In the middle of this inner ear is another membrane (called the basilar membrane) which vibrates. Different parts of the membrane are particularly sensitive to notes of different pitch. Thus, the way in which this basilar membrane vibrates depends on the pitch or frequency of the original sound signal (from about 20 variations per second to almost 20,000 variations per second, the range of sound frequencies the normal young human being can hear).

As the basilar membrane vibrates, small hairs going from this membrane to another nearby membrane which is stationary are forced to bend. This bending of the hairs somehow causes an electrical voltage to appear in the nerve cells which are close to the hair cells. There are about 30,000 such nerve cells in a human ear. Which nerves are excited electrically and the rate at which they are excited depend on the hair bendings, which in turn are determined by the vibration of the basilar membrane.

These electrical signals then travel to the brain through the nervous system. In the brain the original sound is recognized by the nerve signals arriving from the ear.

Research is directed toward an attempt to design devices which can receive the sound signals and generate the appropriate electrical signals to be inserted directly into the nerves leading from the ear to the brain, in other words, a device which replaces man's sensing

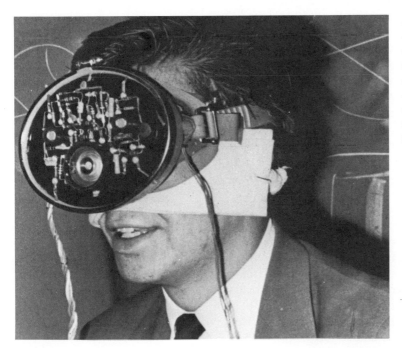

Fig. 9–31.
Newly-developed
amauroscope ("amauro-"
means blind) to "see."
It "catches" light. The light
triggers electrical impulses
that are sent to the brain
through wires attached
to the skull. (Wide
World Photos)

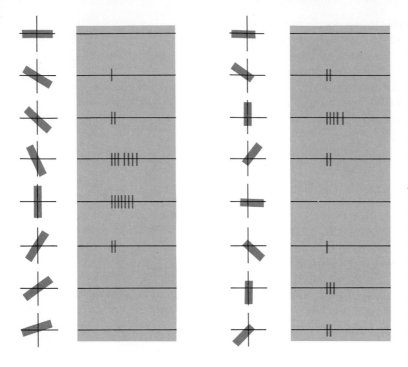

Fig. 9–32.
*Response of visual area
of cat's brain to simple shapes.
In this experiment by physiologists
Hubel and Wiesel, a cat was shown
the various lines indicated at the left.
The electrical activity of single cells in
the brain of the cat was measured. The
spokes indicate the firing (or electrical
discharge) of the cell.*

*Such experiments demonstrate that a
particular nerve cell responds differently
to different shapes. These studies are an
early step toward understanding how
the brain and eye combine to recognize
visual forms, shapes, motions, colors,
and so forth. Similar experiments
show, for example, that certain
brain cells are fired electrically
when the cat sees motion
in particular directions.*
(*Figure from R. L.
Gregory,* Eye and
Brain, *McGraw-Hill
Book Company,
N.Y., 1966.*)

of the sound signals. If such a device can be developed, hearing capability can be restored for people who are totally deaf due to non-functioning of the ear's sound-sensing system.

In order to construct a device of this sort, the engineers and biologists must understand the detailed operation of the sound-sensing system of the ear. How do the electrical signals depend upon the sound waves arriving at the ear? For each possible sound signal, what electrical signal should be delivered to the nervous system? How can this be done by a man-made device which is small enough to be portable (or wearable by the man)? In other words, how can we design a device which matches the compactness, yet remarkably flexible performance of the human ear?

Problems of this nature are even more complex when we turn to a consideration of the eye and the way in which a man sees. Eye glasses are similar to hearing aids in that they assist the man with impaired vision, but they are not useful for the totally blind. Information is sensed in vision by the light passing through the pupil (the opening at the front of the eye) and falling on the retina, which consists of a very large number of light-sensitive cells having varying properties. The electrical signals generated by the retina are transmitted by the optic nerve to the brain. Thus, the design of an artificial eye requires an understanding of what electrical signals result from specific light inputs and in what patterns the human eye operates. For example, how does the eye perceive different shapes, colors, light intensities, and so forth?

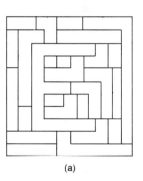

> **Box 9–1**
>
> **PERSONAL IDENTIFICATION**
> (*A human capability machines can't match*)
>
> Have you ever wondered how you identify a particular person? By sight, of course, since humans are good face recognizers, and no two faces are exactly alike. By voice, too, but often we can be fooled since many voices tend to sound alike. Fingerprints uniquely define a person, and toe prints, too. Some muscular-coordination styles are partially helpful, as in gait or in handwriting. Skull structure and tooth geometry, as well as dental work, are highly individualistic; X-rays have been used advantageously to type, classify, and even identify individuals. Cell chromosomes are, of course, unique for each person, but so far we lack the means to read that code adequately. Some bodily characteristics are, for adults, reasonably stable and useful as tags; examples are bone structure, height, skin pigmentation, basic fingernail shape, eye color, and odor. Some occupations and habits stamp us with telltale signs, like skin stains and blemishes from photographic processing, hard manual labor, and smoking. Teeth are deformed by betel-nut chewing. Certain kinds of chemical specificities in addition to blood type and immunological responses can be useful, but our knowledge of these is still relatively primitive. Last, but not least, are all of the mental characteristics and knowledge, quite different for every human being.
>
> Generally, to identify a person positively requires a combination of these characteristics, unless one uses the only proven test, namely, fingerprints.

(a)

(b)

The difficulty of achieving such scientific understanding is a consequence of the remarkable capabilities of the human senses, whether we consider hearing or vision. The human being, for example, has an astonishing ability to recognize patterns. At a large party in a small room, a man can converse with a friend even though the background noise is very great. The man has the capability of picking out a desired message from a noisy background.

The pattern-recognition talent of man is more simply illustrated by the two drawings of Fig. 9–33. If we are told that the first includes a capital letter and the second a numeral, most of us can find these symbols with no difficulty. (They are an E and a 3.) To build equipment which accomplishes this same feat is extraordinarily difficult; we are not even sure what part of this pattern recognition occurs in the eye and what part in the brain. Yet this sort of pattern recognition is exactly the kind of task which must

Fig. 9–33.
Two well-known signs, submerged in "noise." (Colin Cherry, On Human Communication, *M.I.T. Press, Cambridge, Mass., Second Edition, 1966.)*

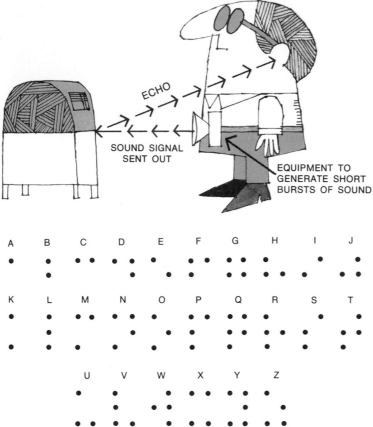

Fig. 9–34.
An
obstacle
detector for
the blind.

ECHO

SOUND SIGNAL
SENT OUT

EQUIPMENT TO
GENERATE SHORT
BURSTS OF SOUND

A B C D E F G H I J

K L M N O P Q R S T

U V W X Y Z

Fig. 9–35.
The Braille alphabet. In this
system, combinations of dots (up to
six per letter) are used to represent
letters and short words. The trained
blind person runs his fingers over
the raised dots. With practice
and experience, most individuals
can achieve reading rates of
30 words/minute (a sighted
person reading 60 pages
per hour is moving at a
rate of about 300
words/minute).

be accomplished if we are to build machines which can automatically convert from spoken words to written text, or machines
which can automatically scan cloud pictures taken from satellites
in order to predict future weather trends.

A normal man receives most of his information about the world
around him through his eyes (for this reason, television is a vastly
more effective means of communication than radio). Because of
this complexity of the sight mechanism, modern technology is making a strong effort to develop devices for the blind which perform
only very specialized tasks. To assist the blind in walking about,
efforts are being made to develop light-weight, simple obstacle
detectors (for example, small sonar sets which send ahead sound
signals and detect any echoes which bounce back from objects—
the system sketched in Fig. 9–34 and similar to that which bats use
to navigate and to capture insects for food).

Another important scientific effort is in the development of
devices to help the blind person to read ordinary printed material
(magazines, newspapers, and books).* The expense of publishing
in Braille (Fig. 9–35) severely limits the variety of reading material

In the United States the scientific research
effort for the blind has really not been a
large national undertaking. Where in the
mid-1960's we spent about $200 each year
per prospective patient on research in heart
disease and cancer, we spent only $1.25 per
prospective patient on research and development of devices for the blind.

Electrical signals representing shape of each letter

Computer—comparison of shape measured with typical letter shapes

Book to be read

Camera (as in TV) which looks at each letter in turn

Electrical signal representing best guess on each letter

Stored sounds for each letter

Sound signal for each letter

Loudspeaker

Fig. 9–36. A reading device for the blind.

s

m

Connection for bellows

Wind box

n

sh

r

Cup-shaped resonator

1 2 3 4 5 6

Inches

Special keys for making sounds: s sh n m r

Fig. 9–37. Von Kemplen's talking machine was powered by bellows (not shown) pumping a wind box, which in turn blew a reed in the throat of the cylindrical cup. Vowels were made by the operator's hand, which partially covered the mouth of the cup, while the nasals and fricatives were made by special attachments. (McGraw-Hill Yearbook of Science and Technology)

available in this medium.* Recent research has included electronic devices to scan letter by letter, then to compare the shape of each letter in a computer with typical letter shapes to find the most probable letter, and then to generate from a loudspeaker or ear-phone an oral statement of the letter (Fig. 9–36). By speeding up the rate at which the successive letters are "spoken" and by train-ing the blind person to recognize shortened versions of each letter, reading speeds comparable to those achieved with Braille can be obtained with ordinary magazines and books.

We have considered only devices to help individuals who are deaf or blind. Equally dramatic advances are being made in other devices. As examples, we might mention:

1. The artificial larynx, or voice-box, to permit a man who has lost this organ to speak again.

*Braille was originally developed for military purposes—in order to communicate written orders at night.

Fig. 9–38. Myoelectric control system for partially paralyzed patient ("myo-" means muscle). The attachment shown allows the patient, with a paralyzed lower arm, to move his "artificial" hand in a normal way. Electrical signals from the nervous system of the patient are used to actuate the electromechanical device. (Dan Antonelli, Rancho Los Amigos Hospital, Downey, California)

2. Artificial arms and legs. Current research is directed to designing limbs which can be moved in a normal way by the usual electrical signals from the brain (Fig. 9–38).

3. The heart pacemaker, a device inserted under the skin of the stomach with wires going to the heart. The device generates electrical signals which regularly shock the heart and thereby cause the heart to beat, for patients in which beating is impaired or too slow or irregular. By 1970, pacemakers had been placed in more than 100,000 people in the United States. New devices were being used at the rate of 15,000 a year. Once the pacemaker is in place, life expectancy is about 90% of what it would be if the man had never had the heart trouble at all.

4. Arterial replacements. The replacement of a diseased or weakened section of the aorta with an artificial artery is now a relatively common operation.

The successful development of each of these examples depends upon a scientific understanding of the detailed way in which the human part operates. The technological possibility of building replacement elements for the human body thus stimulates the medical researcher to learn in greater detail about the biological phenomena. Thus, our knowledge about science grows in parallel with our ability to build more and more sophisticated man-made devices.

7 | MATCHING TECHNOLOGY TO MAN

Throughout this chapter, our emphasis has been on using technology for the benefit of man. When we speak of the rest of the twentieth century as the *age of technology*, we mean that this technology to serve man is appearing in more and more of our lives. Man's work and recreation, his communication and travel, his education and enjoyment of the arts—in all directions, technology properly used can enrich life.

The key to success here is in the idea of technology *properly used*. What is implied by this phrase?

Meeting Man's Needs

First, we must know man's needs.

The subject of nutrition is an obviously important example where we need to know the needs. Discussions of the world population problem usually dramatize the problems of famine, particularly in Asia. Any planning for new sources of food or new methods of food distribution must be based on knowledge of man's daily requirements to assure health and the ability to work and live normally. Yet our scientific knowledge today does not even include the number of calories an adult must eat each day to avoid losing weight. There is no general understanding in detail of why some people are overweight.

The problem is difficult because nutritional needs are apparently highly individualistic. In other words, what one person needs may be very inappropriate for another. In England, the disease of rickets was controlled by giving babies vitamin D in their food. Soon after the program started, it was found that certain babies

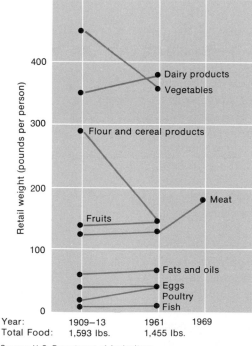

Year: 1909–13   1961   1969
Total Food: 1,593 lbs.   1,455 lbs.

Source: U.S. Department of Agriculture

Fig. 9–39.
Changes during this century in the foods Americans eat. During the 50 years shown, the total quantity of food per year dropped by 10%.

were dying from kidney trouble, traced to too much vitamin D. When the vitamin D was decreased, rickets reappeared in other children. The solution lies in an individual nutritional program. Just as in the overweight problem, each person must be considered as an individual.

Furthermore, nutritional needs must be determined very early. We now know that the maximum size of various organs of the body is determined before birth. The brain reaches its full number of cells by the age of two. Experiments with rats suggest strongly that a shortage of milk during the first few months of human life could lead to a small number of brain cells and, possibly, changes in mental behavior. Scientific experiments on people have not been performed, but there does seem to be some relation between nutrition in infancy and mental development. There certainly is a relation between nutrition in early childhood and the way the adult later produces fat. The overnourished child develops a lifelong desire for extra food.

In other words, the problems of nutrition are an example of an important area where the possible benefits of technology are severely limited by our inadequate knowledge about the needs of man. We continually seek new methods of food production and distribution. These developments can never provide enough food to prevent starvation and malnutrition unless we learn how to measure accurately minimum human needs and then base our planning on this system knowledge.

Fitting Man's Characteristics

To use technology properly, we must also design the system for the user. The machine must take into account man's capabilities and limitations, his likes and dislikes. The problem displayed in Fig. 9–40 is a simple example of this entire area. Here are pictured the tops of four different cooking ranges, each with four burners labeled A, B, C, and D. Each burner is controlled by the knob with a similar label. For each of the four models, tests have been made to determine how easily people can learn to associate the proper control with a particular burner. The results are startling in their differences. System I (which happens to be the arrangement seldom used in commercial products) proves to be very much simpler to learn than the others, and the complexity increases sharply from I to IV in the diagram.

Figure 9–41 shows another example of technology which does not seem to be matched to the human user. The top part shows the way the manufacturer designed the heat controls on an electric range, while the bottom portion shows the way they should be designed to match the normal expectations of a housewife (as determined by surveys of women). The housewife expects the heat to increase as one moves from left to right, she anticipates 3 to

Fig. 9–40.
Models of four stoves.
(From Man-Machine
Engineering by Alphonse
Chapanis. Copyright 1965
by Wadsworth Publishing Company,
Inc. Reprinted by permission of
the publisher, Brooks/Cole
Publishing Company,
Monterey,
California.)

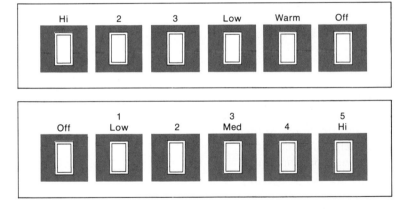

Fig. 9–41.
Pushbuttons for an
electric stove as the
manufacturer originally
designed them (above), and
the stereotypes which housewives
had about these pushbuttons (below).
(From Man-Machine Engineering
by Alphonse Chapanis. Copyright 1965
by Wadsworth Publishing
Company, Inc. Reprinted by
permission of the publisher,
Brooks/Cole Publishing
Company, Monterey,
California.)

represent more heat than 2, and she assumes "Warm" means more heat than "Low." In other words, the original design contains a variety of sources of significant or even dangerous errors in use of the technological device.

Figure 9–42 is a final example of the contrast between poorly designed devices and devices matched to human characteristics. A micrometer is used for precise measurements of thicknesses of a solid substance. Each micrometer is set at a reading of 0.398 inch. The top unit, which was standard in laboratories until very recently, is read with the two dials in combination, with the dial around the rotating handle used for the fine interpolation between divisions on the principal scale (shown as 0, 1, 2, 3, 4).

Fig. 9–42.
Micrometers
(a) Conventional
design. (b) Designed
for easy readability. (From
Man-Machine Engineering *by*
Alphonse Chapanis. © *1965 by*
Wadsworth Publishing
Company, Inc., Belmont,
California 94002.
Reprinted by
permission of the
publisher.)

These examples emphasize that the usefulness and value of technology depend not only on the ability of the device to do something which is meaningful for man, but also on design so that the device is suitable for man's use. Fortunately, in recent years manufacturers have been developing a growing appreciation of the importance of this part of technology, and we now frequently see extensive consumer testing of new devices in order to ensure satisfactory performance. There has been a major effort to improve highway road signs, but it is still relatively easy to find signs indicating a turn-off after it is too late to move safely into the turning lane. New railroad cars and airplanes are being designed from a knowledge of typical physical size of the individual American traveler. Laser obstacle detectors to assist the blind to move about are being designed on the basis of research on the needs and wishes of the blind.

To conclude this section, one example illustrates the potential seriousness of poor design. Figure 9–43 shows a possible factor in the disastrous airplane collision over the Grand Canyon in 1956—a crash which took 128 lives. As indicated by the figure, visibility from the cockpit of many modern airplanes is severely limited,

Fig. 9–43.
*Poor visibility to the rear
and below probably caused the
Grand Canyon crash. Although the
exact positions of the two aircraft
are not known, this drawing shows
how the limited fields of vision from
the converging DC-7 (behind) and
Super-Constellation could have
prevented each pilot from seeing
the other plane until too late.
The triangles show the approximate
field of vision from each airliner.
(From Man-Machine Engineering
by Alphonse Chapanis. Copyright
1965 by Wadsworth Publishing
Company, Inc. Reprinted by
permission of the publisher,
Brooks/Cole Publishing
Company, Monterey,
California.)*

and it is entirely possible that the pilots could not see the other plane until just before the impact.

As technology becomes more elegant and more of everyone's life is directly concerned with technological systems, the best design of equipment becomes more significant. The automated highways and trains of the future must be both entirely reliable and excellently matched to the characteristics of the human users.

Controlling Technology's Side Effects

Finally, if technology is *properly used*, the side effects must be anticipated. We are all familiar with the new, expensive, eight-lane highway which suddenly ends at the city line. Cars move into the city at a delightful speed. Indeed, the trip is so easy that many more people now drive. But there is nowhere for the extra cars to go (or even to park). This is an example of planning we might call ill advised. Unfortunately, the problem is often much more serious and difficult. With the rapid introduction of new technology, we can expect dangerous side effects unless a major effort is made to carefully plan ahead. We have already seen such effects in the thermal pollution from nuclear power plants. It was only recently that nuclear energy was widely heralded as bringing to an end the air pollution caused by coal and oil plants. Now, only a few years later, we find serious ecological changes in the waters near nuclear stations—even massive fish kills.

The problem of predicting side effects of new developments is called *technology assessment*. In an assessment, we attempt to anticipate the effects not only on the system in which the technology is used, but also on related systems. For example, the proposal has often been made to require all taxis in a city to be electric cars. The direct hope is to reduce air pollution.

What are some obvious side effects? First, the cost of taxi rides would rise. We should estimate the increase, then predict the number of people who would (as a result) bring their own cars into the city. The electric energy used to recharge the taxi batteries would have to be generated and distributed. What changes would this require in the electric utility? Would electricity rates rise? If so, how many businesses would move out of the city, with resulting unemployment? How serious would be the traffic problem because of stalled taxis? Where would battery-recharging stations be located? Who would man these stations? How would traffic flow be affected? Would people who use taxis regularly move to the suburbs?

Clearly, we can go on and on. The new technology which at first seems so attractive actually has far-reaching effects. Unfortunately, as the ecologists have emphasized, such effects are often obvious only after they are irreversible.

8 | FINAL COMMENT

In this chapter, we have considered briefly a few of the problems associated with the human use of technology. From these discussions, it is apparent that we really cannot separate science and technology from many other fields:

Biology (for its study of man's abilities and limitations).
Psychology and behavioral science (for the study of man's attitudes, likes, learning ability, etc).
Social sciences (for the evaluation of the effects of technology).

In the first nine chapters we have met a variety of the concepts which are the foundation of modern technology. In the last six chapters, we will sharpen our focus. We will concentrate on one product of modern technology: the computer. From this study, we will gain an appreciation of how we can assemble an astonishingly complex man-made system on the basis of very simple rules, how we can utilize technology to expand man's scope, and how modern technology can be designed to serve man's needs.

It is appropriate that we select the computer for this focus. The computer is a monumental achievement of twentieth-century man. It illustrates dramatically the crucial importance of a logical, quantitative approach to system understanding. Finally, the computer, still in its infancy, will profoundly influence individual lives as well as the complete social structure during the remainder of this century. In other words, the computer is central to the interrelationship among man, technology, and society.

Questions for Study and Discussion

1. What are some of the criteria we use when we evaluate the way in which a device is designed for a specific human purpose? Give at least two examples of such devices.

2. In the area of locomotion, what is the major advantage which the machine has over man?

3. In the area of communication, what advantages does man have over the machine?

4. In the area of control, what is the major advantage a machine has over man?

5. What are the major sources of error in electronic transmission of a binary signal?

6. Describe the "cocktail party" effect in communications.

7. The word spelling code used by the United States Armed Forces during World War II had the following word alphabet:

Able	Jig	Sugar
Baker	King	Tare
Charlie	Love	Uncle
Dog	Mike	Victor
Easy	Nan	William
Fox	Oboe	X-ray
George	Peter	Yoke
How	Queen	Zebra
Item	Roger	

What advantages does the ICAO code discussed in the chapter have over this one?

8. Below is a block diagram of a human control system. Using the diagram, explain how a person maintains balance when riding a bicycle, a skateboard, a surfboard, or backing his automobile out of the driveway.

9. Give two examples in which a technological system is superior to a human. In each example point out the specific parts of the system in which the machine is superior.

10. Give two examples in which the human is a vital part of a technological system. In each example point out the specific task which can be done better by the human.

11. What is the major purpose of the automatic error correction code described in Section 3?

12. Section 5 lists many quantities which man cannot sense directly. Prepare a list of quantities which man can sense and for which he seems to need no help from technology, as well as a list of quantities which man can sense, but for which he sometimes uses technology to aid his sensing.

INDEX